William Wright

The Oil Regions of Pennsylvania

Showing
Where Petroleum Is Found; How It Is Obtained and at What Cost
With
Hints for Whom It May Concern

Elibron Classics
www.elibron.com

Elibron Classics series.

ISBN 1-4021-6448-3 (paperback)
ISBN 1-4212-8106-6 (hardcover)

This Elibron Classics Replica Edition is an unabridged facsimile of the edition published in 1865 by Harper & Brothers, New York.

THE OIL REGIONS

OF

PENNSYLVANIA.

SHOWING

WHERE PETROLEUM IS FOUND; HOW IT IS OBTAINED, AND AT WHAT COST.

WITH

Hints for Whom it May Concern.

BY

WILLIAM WRIGHT.

NEW YORK:
HARPER & BROTHERS, PUBLISHERS,
FRANKLIN SQUARE.
1865.

PREFACE.

In the latter part of March, I left my home for the South-West, designing to pass through the heart of West-Virginia, Eastern Kentucky, Eastern Tennessee, and thence down the Tennessee valley—directly as a correspondent of the New-York *Times;* indirectly with a view of collecting materials for a volume on the Border States, their soils, minerals, climate, water-power, social condition, etc., for the guidance of those who might desire to migrate thither after the war.

On the way, I proposed to spend a few days in the oil regions of Venango county, Pennsylvania, and afterward visit those of West-Virginia, supposing that a week would probably suffice to do both all needed justice.

In accordance with this plan, I walked, jumped, or waded the valley of Oil Creek, from Titusville to Oil City, collecting what facts and observations I could during the three days consumed in the passage.

Arrived at the journey's end, I found a discordant, contradictory mass of facts and figures on my memorandum-book; and came to the conclusion that, whatever I knew the first day, I knew much less the second, and nothing at all the third. Further, that no person outside of Petrolia, and very few in it, were in a much more enviable condition of mind on the subject, if they would own up to the truth.

After deliberating afresh, I formed the resolution of visiting every producing well in that county; gathering from men, who were supposed to have no interest in misrepresenting, its actual yield; comparing the figures with those given by officials and neighbors, and out of the whole endeavoring to ascertain the truth, as nearly as might be. At the same time, to "bore," and "ream," and "pump" every practical man for the results of his observations, if so be it were possible to arrive at *one* general law or conclusion respecting the oil regions.

The residue of March and nearly three weeks in April were faithfully devoted to this object. The distance traversed on foot was fully two hundred miles—how agreeable the trip, will be easily inferred from what follows.

An interest, having more than $100,000,000 of *bona fide* capital invested in it, had until then never received more attention than could be given it in newspaper correspondence or a magazine article. The financial aspect of it had not even been *scratched*. Indeed, honest writers seemed to avoid reference to it, except in the most general terms, as if it were going beyond their depth. Of course, the Oily Gammons of the press, who had been hired as *claqueurs* at a theatre, applauded every thing. That was their vocation!

In the following pages I have described the processes of boring the wells, of repairing them after getting out of order, and of refining the oil. I have entered somewhat minutely into the physical formation of the country—a topic which had been almost overlooked, and on some points of which, I hope to have thrown out some valuable ideas for the first time. When adopting the views there presented, I had not perused the Geological Report of Prof. Rogers; and it is highly gratifying to find that in the main features of the argument advanced, I am fully borne out by that eminent name.

But it is to the statistical and financial discussions that I desire principally to direct attention. Those chapters will be read by large numbers who are eagerly in quest of the information therein contained. The facts and figures now given to the public for the first time, together with the modes of taking in over-smart, shrewd, keen, knowing Eastern people, will *tell.*

Petrolia needed a searching examination and a scathing exposure; it has got both. Yet let me not be misunderstood. Underneath a system of falsehood and fraud, that might almost be termed *magnificent*, there is a great basis of fact, which needs to be presented in its true light; needs to be protected from the misrepresentations of its own pretended friends, who would have ruined it long since if it had not possessed genuine worth of a high order. It is to censure what is worth censuring; to strip off and expose what is false and deceptive; to denounce the cruelty, the lying, the roguery, the abject selfishness of many, that I have for the time being turned aside from my original object to prepare these sheets for the press. I have aimed to state the truth without calumny or prejudice; to express it clearly and forcibly; to be as thorough as it was possible within moderate limits. How well or how ill these objects have been accomplished, the reader will judge for himself.

It is with a feeling of gratitude that I acknowledge the courtesy, in imparting information, of Messrs W. H. L. Smith, of Corry; A. Morrell and Robert B. Gamble, of Titusville; Edward Fox, of Petroleum Centre; Wm. Boniface, of Rouseville; T. S. Truaire and C. B. Bliss, of Oil City; Thomas R. Hennon, of Tideoute; Col. McClure, then of Plumer; George S. Siggins, of Howe; and many others, whose names I do not now recall.

Since this volume went to press, reports have been received to the effect that the United States well on Pithole Creek has increased its flow to nine hundred or a thousand barrels per day; other wells in that locality are also said to have improved. On the other hand, certain wells, as the Jersey, on Oil Creek, have fallen off or dried up altogether. No doubt, however, the summer product of petroleum in Venango county is considerably larger than that of March and April. By referring to the last chapter but one, it will be seen that a margin of about two thousand barrels per day has been allowed for this increase.

W. W.

PATERSON, N. J., May, 1865.

CONTENTS.

THE OIL REGIONS OF PENNSYLVANIA.

CHAPTER I.

PHYSICAL FEATURES AND GEOLOGY OF THE COUNTRY.

THE physical features of the oil region of Pennsylvania are simple, easily understood, and full of interest. The most productive portion of it consists of an irregular quadrangle, each of its sides being from fifteen to twenty-five miles in length, and its axial line nearly corresponding with the course of Oil Creek between Titusville and Oil City. As far as known, it is limited almost exclusively to the Alleghany River valley and a section of its north-western slope, the principal streams which enter it from that direction having an average fall of about twenty feet to the mile, while that great artery of Western Pennsylvania, between Warren and Pittsburgh, has an average fall of one foot to the mile. At Franklin, the mouth of French Creek, it is nearly eight hundred feet above tide-water, or two hundred and forty feet higher than the surface of Lake Erie. The lake shore proper is a comparatively narrow strip of land, descending from the range of heights which separate it from the Alleghany slope, by frequent and abrupt terraces, to the water. In the opposite direction the descent toward the south-east is for a long distance so gentle as scarcely to be noticeable along the water-courses. The country is nearly level, or rolled up into hills of moderate elevation and easy ascent. Such is its general appearance immediately back of the oil region along the line of the Atlantic and Great Western Railway.

On leaving Corry for Titusville, the passenger is conveyed up a rather steep incline for a few miles, on passing which the railroad traverses the summit, a table-land with occasional ridges of very moderate height, or round hillocks formed of *drift*—that is, sand and boulders, ground, rounded, and more or less polished. By and by the road enters a slight depression, which increases till it becomes a ravine, and finally the bottom of a noisy stream, with banks twenty, fifty, and finally one hundred and fifty feet high, at which it reaches Titusville. The railroad extends down the valley seven or eight miles further, the heights on each side becoming more lofty and precipitous, until, at Shaffer's, they reach two hundred and fifty feet, and at Oil City, twelve miles lower, nearly four hundred feet above the Alleghany River, the difference in their summits being inconsiderable.

The approaches to that river by the railroad from Meadville to Franklin, and by that from Corry to Irvine and Warren, are still more simple than on the route already described, there being no elevations to be climbed before reaching the descent; but a gentle, downward grade for the entire distance. Starting from a nearly level country, however, the same phenomenon is perceptible as on the Titusville road, namely, heights gradually increasing in altitude and abruptness, until the common objective point (the Alleghany valley) has been reached, where they are found at about the usual elevation of four hundred feet.

Starting from that river in the direction of the Alleghany Mountains, along any of the tributary streams which flow through that part of the country, an exact counterpart to this formation will be found, for the twenty, thirty, or forty miles immediately east of the Alleghany.

Not alone in the oil regions of Pennsylvania will this observation be found to apply. The same physical features are noticeable in Western New-York, West-Virginia, Eastern and Southern Ohio and Indiana, and Northern Kentucky—in fact, throughout the entire prairie States, along both sides of the Mississippi and the Missouri. Whatever may have been the cause, it is evident that agencies essentially alike have been at work in producing the effects visible to-day.

The observant visitor will notice, further, that all this is accompanied by hardly a perceptible inclination or "dip" of the rocks, which, as high up and as far down as he can trace them, are disposed in beautiful horizontal lines, gray, yellow, or brown sandstones alternating with blue, red, or brown shales, in regular and frequent succession. At no point will he detect "faults" or dislocations in the arrangement of these, whether on the surface or beneath, nor will he find the inclination at any point exceeding five degrees. If he has read or observed carefully the structure of mountains elsewhere, he will remember that the rocks are tilted up at various angles with the horizon, in some cases almost ninety degrees; while at every few paces he can discern marks of disturbance in their situa tion with respect to each other. He will begin, perhaps, to reason within himself how very improbable it must be that those heights could have been *individually* upheaved, inasmuch as they evince no signs of such action, having no inclination save that south-eastern or south-western dip common to all the formations from Lake Ontario to Kentucky and Tennessee.

If he climb the heights, the mystery at first may increase; for while toward the principal streams the bluffs are usually abrupt, precipitous, pyramidal, or ridgy; on looking in the opposite direction, he will observe the same table-lands, with gentle elevations formed of drift here and there, such as he encountered between Corry and Titusville. He can hardly fail to suspect the truth, that the river-beds, with all their tributaries and the bottoms of ravines, have been, in the course of ages, *scooped out of very slight original depressions by the action of water*, which has dug out these vast channels to their present depth, and is still engaged in deepening and widening them.

This is the view taken by all geologists of eminence, who have examined that portion of the State. Professor Wm. D. Rogers, in his admirable Geological Report, says: "The Alleghany flows through a deep and narrow trench, excavated in the north-west plateau, and within the coal-basin of the State. . . . It enters the north-western margin of the coal-basin at Warren." Referring to the remarkably irregular course of that river and the Ohio through Pennsylvania, he further observes: "It is evident that while the main discharge of the eroding wave was south-westward, one large influx of eroding waters swept north-westward from the Apalachian Mountains, and another south-eastward from the region of the lakes."

The expression "eroding wave" must not, however, be understood to imply that, by a single deluge, whether lasting a day or a twelvemonth, the troughs in which those rivers now run could have been hollowed out. No single accumulation of waters, collected on the Apalachian Mountains or in the lake basin, would be sufficient to excavate those valleys to the depth of a single inch, much less hundreds of feet. The joint action of time and water was necessary to "write those wrinkles on the brow" of that Piedmont section; and of the former it would be silly to assign less than myriads of our years. It is true that in an earlier stage the denuding process would go on more rapidly, inasmuch as the mouth of the Mississippi was several hundreds of miles higher up than at present; while the Ohio and its tributaries, flowing down from the table-lands, would rush forward and abrade their bottoms and banks at a rate which can hardly be understood at present. The hard limestone bed crossing that river at Louisville has prolonged resistance till this day; but that is also destined to wear away, together with the softer rocks above; the fall or rapid thus, step by step, retreating to Pittsburgh, unless prevented by artificial obstructions.

Along the Ohio, as it passes between Pennsylvania, West-Virginia, and Kentucky, on one side, and Ohio and Indiana on the other, it may be observed that while there is substantially the same horizontal stratification, with abrasion of the uplands, the action there has been on a still larger scale than further northward. The causes of this difference are unquestionably to be found in the circumstances of the larger bodies of waters at work lower down, and the beds of rocks forming the upper series along the Ohio consisting of soft shales, which readily dissolve, and, in places, are so constantly crumbling away as to form little streams of gravel, which become beds of stiff clay or mud upon reaching the bottoms. This is carried off in immense quantities every year, on the melting of the snows and after rains, to the Ohio, the Mississippi, and the Gulf, to form fresh additions to the area ot Louisiana, or some other of the Cotton States. For the whole of the sugar region, and at least one half of the cotton, rice, and tobacco zones of the South consist of alluviums, carried down the eastern and western slopes of the Apalachian and the Rocky Mountain systems, along river-courses, whose beds have been carved tens and hundreds of feet below what was once the general level of the country.

In the oil region of Pennsylvania, the uppermost rock in the series (vespertine) is a hard, gray conglomerate, which resists the action of water until undermined by the dissolution of the soft shales underneath. When this has been brought about, down tumble huge masses of the vespertine, in uniform and often quadrangular blocks, which in turn become the natural protectors of the bases and sides of the bluff, as they had been of its summit. But for these efficient safeguards, it is reasonable to infer that the channels of the Alleghany and some parts of the Ohio would have been as wide as they are found elsewhere, particularly near the mouth of the latter, where the country for miles on each side has been washed away.

This explanation of the origin of those valleys finds abundant confirmation at every bend in the streams, particularly the large ones. In proportion to the sharpness of their curves is the precipitous character of the hills forming their exterior lines; while on the opposite side, the heights slope gently and gracefully to the water. Almost the only fields cultivated on the slopes are those headlands around which the river describes a semicircle. On rivers flowing through alluvial countries, as the Potomac and the James, the same rule will hold good, as to the areas embraced within their several bends or loops, while the opposite shore will invariably be found lofty and abrupt, the channels keeping well outward, as if desirous of increasing their circuit, instead of proceeding "on interior lines" or cut-offs. The phenomenon is thus explained: Like all other bodies, water has a tendency to move in straight lines, as have all the objects which move upon or are carried down by it. At every bend, for example, the raft is naturally thrown somewhat out of the middle toward the convex side of the river, and the shore there is struck with all the weight and momentum of the current, until it is undermined, when the portions lying highest up fall down and are gradually ground up and their particles carried off, making room for the water to renew its action upon the bluffs. Hence the two processes of lengthening and deepening the stream may be said to go on *pari passu*. At the principal bends, the river, which at one time flowed on a bed hundreds of feet higher, also described hundreds of yards of a shorter course than it now does. As it deepened in one direction, it struck outward in another. During century after century, it thus kept forsaking its old channel and entering one more crooked, rounding the headlands, which were afterward covered with surface soil washed down from above, or made by vegetation on the spot. Every mile or two along the Alleghany, Oil Creek, French Creek, etc., these additional evidences of erosion or denudation become manifest in the formation of the river-beds. The unavoidable inferences are, that the depositing of oil, or the constituents composing it, has had no possible relation to existing river systems; that curves, headlands, slopes, table-lands, and bottoms, have no connection, *as such*, with the finding of petroleum; the only cause why it was discovered in the last-named being, that that which was stored in the first sand-rock beneath the surface could more readily find its way to it in localities where the deposit had been approached to within short distances. It was only because it could come more easily to the surface in springs along the river-bottoms than on the uplands, that it was originally discovered in the former; but on the latter, the veins in the second or third sand-rocks, which are unable to manifest themselves above ground, are quite as likely to be reached as on the former, the only drawbacks being the greater cost of boring and of pumping the wells afterward.

Though rather foreign to the subject discussed in this work, I may add, that the views advanced above explain the cause why "cut-off" canals, as those at Vicksburgh and Dutch Gap, have been unsuccessful. They were dug at points too far down-stream, with their upper extremities some distance below the bluffs, against which the current had struck and been deflected toward the opposite bank. Had General Butler's famous work taken "the line of beauty," its northern terminus opening against the channel as it struck the bank, there could hardly be a doubt of its success.

If the length of time required to effect such changes on the earth's surface be bewildering, our wonder will not subside on inspecting more closely the several strata between the uplands and the lowest point yet reached by the drillers, a range of about fourteen hundred feet. Marvels, indeed, never cease in Petrolia, whether we regard it in its natural aspect. or its lately developed condition. If we take our stand on any of the elevations along Oil Creek, French Creek, the Brokenstraw, or any other tributary of the Alleghany, in that quarter, in the masses of gray, brown, or flaggy sandstones, even on the conglomerates forming the uppermost rock in the series, we shall discover multitudes of the impressions of what once was vegetable or animal life—shells univalve and bivalve, shells of all shapes and sizes, rough and smooth casts of shells. The plants belong largely to a species known among naturalists as fucoids. The organic remains of the shell-fishes indicate that they lived and moved in salt, not fresh, water; that they died a natural death, the consequence of ordinary decay, closing up the doors and windows of their abodes, and quietly dropping to what had been their couch ere it became their cemetery—the ocean-bed—as composedly as Cæsar wrapped himself up in his toga, at the base of Pompey's statue. Nothing out of the natural course, no marks of violence are perceptible in the circumstances attending their dissolution. No deluge appears to have swept them away five hundred or one thousand miles from their native settlements to end their existence on the tops of the mountains.

> "Once, in the flight of ages past,
> There lived a *crab*; and who was he?
> Mortal, where'er thy lot be cast,
> That *crab* resembled THEE!"

This testimony is not confined to one rock or series in the several geological formations of Western Pennsylvania. In the numerous layers of sandstone, slate, shale, clay, and soap-stone, evidences of mineral and vegetable life may be found as low down as the beds of the streams, and hundreds of feet beyond. At different points on the railroad between Meadville and Franklin, the rock-cuttings pass through what once were dense forests of tall trees, the petrified ends of their stems projecting through the soft layers, which are beautifully bent around them, the petrifactions lying in places as closely together as they could have fallen.* Their diameters range from ten inches to three feet. The pores of the bark are still visible, as is its general arrangement on the outside of the quondam trees. The grooves on the exterior, the layers of the interior, with several of the lines radiating from the heart, are still observable. These petrifactions may be found at various depths, from ten to forty feet below what is *now* the surface of the height, but which was hundreds of feet below it previous to the eroding action of the waters. Shell-marks in abundance are discoverable in the lowest of those rocks, as in all the intermediate strata. Mr. A. Morrell, now of Titusville, an experienced operator and a careful observer, assures me that the sand-pump has brought up, from the depth of four hundred and eighty-five feet, specimens of petrified shell-fish, which are now in his possession, having been obtained in a layer of hard, fossiliferous limestone, termed by most operators, "the third sand-rock." Here we have an aggregate thickness of fully seven hundred and fifty feet, containing probably twenty distinct layers, in most or all of which the remains of vegetable and animal life are discovered. Should it turn out that the Niagara limestone underlies that re-

* The remains of a mastodon are said to have been discovered in that vicinity, while the railroad was being graded.

gion, there is no telling how many hundreds of feet lower such remains may yet be discovered.

The groups of sandstones *above* the river-beds differ somewhat in various localities, one of a coarser texture being replaced by a finer, or the opposite, a pure sand-rock, by one more or less argillaceous, a gray by a yellow-colored conglomerate, containing pebbles of quartz, sienite, gneiss, etc. In some places they afford excellent materials for building purposes; in others they are made into grindstones; superior flag-stones are to be met with everywhere. But such are mere local variations, brought about by the deposition of the several materials of which the layers are composed. In general, it will be found that they belong to rocks which terminate toward the north in the State of New-York, and, in South-western Pennsyvania and West-Virginia, are found hundreds of feet below the surface, the several coal-formations, with the intermediate layers, being there superimposed on these formations. Counting *downward*, in Venango county, the following is the succession which most frequently occurs:

1st. *Vespertine conglomerate and sandstone.* This is a white, gray, or yellowish rock, varying in texture, and alternating with a coarse, silicious conglomerate, or with dark-blue and olive-colored slates. In places it contains beds of dark, carbonaceous shales, with thin seams of coal. Among its organic remains are numerous fragments of coal-plants. In some parts of the State this series is found of immense thickness—as much as twenty-six hundred feet, according to Professor Rogers, near the Susquehanna River. In Venango county I have not observed it more than one one hundred feet deep, and seldom more than fifty. "This group has a wide distribution in Pennsylvania, encircling as with a sort of girdle, all the coal-fields, both anthracite and bituminous. . . . It undergoes gradual but important changes of type, growing thinner, and assuming a finer and finer texture in its materials, as it spreads westward. Its orographic position is on the ridges and external escarpments of the table-lands, which enclose or support the coal-fields; but, except in the north-western district, it does not immediately adjoin the conglomerates and sandstones of the coal-measures. . . In this north-west belt, and along the north side of the State, it is a somewhat argillaceous, micaceous sandstone. . . An absence of fossils, and a close assimilation of the vespertine and umbral series render it difficult to define their common boundary. . . Around Warren, this group of rocks, reposing directly on the vergent series, and overlaid by the seral white sandstone, consists of four members—the lowest a group of thin-bedded sandstones and olive-gray shales, the sandstone containing a perpendicular, bifurcating, stem-like fossil. The second is a massive quartzose conglomerate of smooth, ovoid pebbles, about ten feet thick. The third is a thick mass of olive-gray shale and thin-bedded sandstones, about one hundred and seventy-five feet. The fourth, or uppermost, is a fossiliferous gray sandstone, from ten to fifteen feet. . . The vespertine conglomerate caps the hills north-west of the Alleghany River. It is often mistaken for the *seral* conglomerate and sandstone of the coal measures." (ROGERS.)

2d. *The vergent series.* This consists of a rather fine-grained gray sandstone, the layers parted by thin alternating bands of shale. According to the authority just quoted, it abounds with remains of marine vegetation. In Huntington county it attains a thickness of seventeen hundred feet. In Venango county it comes down to the river bottoms, a distance of nearly three hundred feet below the vespertine rocks; how far beneath the valleys is unknown. Toward the west this series has a wide extension, spreading out in Ohio, in Kentucky, and even Middle Tennessee, as well as stretching through West-Virginia and East-Tennessee. Some geologists have classified this series as follows, counting downward: Slate and flaggy sandstone. Two layers of hard, silicious sandstone. Beds of thin, pebbly rock. In places, a stratum of yellow sandstone. Several beds of gray sandstone, of more or less thickness. Two or three thin layers of sandstone, with shells. Beds of shale, usually very dark, alternate between each of the above-mentioned layers.

This second series comprises what are known in New-York as the Chemung and the Portage groups. They belong to the class denominated *palæozoic* rocks, because containing the most ancient remains of animal and vegetable life yet discovered, stretching all the way between the gneissic formations beneath, and the lowest of the coal-deposits above. Sometimes they are denominated, "fossiliferous," "sedimentary," or "secondary" rocks. "In Pennsylvania," says Rogers, "this class have been deposited during all the four earliest periods of the great European divisions, namely, the cambrian, the silurian, the devonian, and the carboniferous. No traces of the fifth or permian group have yet been discovered in North-America. . . . The prolonged succession of sedimentary action ceased with the close of the cambrian system, being terminated by the upheaval of the ocean, in whose broad bed, and around whose margin these deposits had been collected." The aggregate thickness of all the rocks belonging to this class, measured at their greatest depths, is not less than thirty-five thousand feet.

As mention has been made of the coal-beds found in that vicinity, and as the coal and petroleum formations are generally believed to be intimately related to each other, both being largely carbonaceous, it may be proper to remark here, that the oil region is bounded by the coal-fields, on the south and the east. On both sides of the Alleghany, near Franklin, coal crops out, in a thin bed, and of a rather poor quality, along the summits of the hill, from two hundred and fifty to three hundred and fifty feet above water. Farther down the river it occurs at a lower elevation, until on reaching Pittsburgh the deepest beds are found below the water-courses. In some localities near Franklin, the coal-vein appears to have filled a cup-like depression once existing on the surface where it was formed, the bed now dipping downward, and afterward rising to its former level. In that neighborhood the coal-bed may be traced as a dark, even line, extending along the highest eminences, while at the foot of the bluffs operations at boring have been going on. Toward the east the coal-fields are more distant; but their margin in that direction has not been clearly traced and defined. Owing to the absence of large streams and broad bottoms, operations at searching for petroleum have also been greatly restricted east of the Alleghany. The nearest practicable point at which coal can be reached is on the line of the Philadelphia and Erie Railroad, above Warren.

As far as I am aware, no actual test has yet been made, or if made, none has succeeded, in boring for oil through the coal-measures. The experiment is one that ought to be instituted, its results having something to do with verifying or disproving the theory as to the genesis of petroleum commonly entertained. By some it is stoutly maintained that the one is never found immediately above or below the other; hence, that it is idle to expect oil on coal-lands. Others assert that they have, in West-Virginia, passed through the coal, and found oil beneath. It devolves upon the owners of coal-lands to prove the contiguous existence of the two, or else be satisfied with lower prices than they are now demanding.

At all events, the two formations in Pennsylvania are *geologically* separated by great distances. Between the coal-beds and the first sand-rock where oil appears, at least four hundred feet intervene, and at least eight hundred feet between the coal and the third sand-rock, from which the largest yield has been derived. As to the possibility of the one being supplied from the other, some further remarks are offered elsewhere.

Descending to the formations *below* the river-bottoms, as traversed by the drill, and brought up (as mud) by the sand-pump, it becomes necessary to take the "records" of wells as kept by operators of more or less acquaintance with the subject. Mr. Morrell gives the following as the series found on the Watson flats below Titusville, in sinking five hundred feet:

First. Schist rock, with mica and a little hornblende. This is what the drillers usually term "the first sand."

Second. A very soft, silicious rock, resembling soapstone. Feels very soft and greasy at first, and has a nearly white color; but hardens, and becomes bluish after a time.

Third. A hard transition rock.

Fourth. Fossiliferous limestone, containing fissures oı

caverns in which oil is found. This is usually termed "the second sand-rock;" and as it contains sand in various proportions, it will be so termed in this work.

Mr. R. B. Gamble, also of Titusville, who is superintendent of the Pennyslvania Oil Creek Company's Deep well, (now fully twelve hundred feet below the level of the Watson flats, and still in progress,) furnishes the following as the results of his observations:

The first sand-rock commonly occurs there between one hundred and fifty and one hundred and sixty feet down, and is about sixteen feet thick. Drillers then bore through different layers of slate and soap-stone, (not the common variety of the latter,) in which the tools often stick fast. Between four hundred and eighty and four hundred and ninety feet, they strike a pebbly bed, often mixed with slate, and usually five feet thick. What is known as the second rock occurs about four hundred feet down, and is between fifty and sixty feet thick. In this bed, oil is most commonly obtained in that locality. In this particular instance, the company decided to keep on boring for the purpose of making an experiment, with the annexed result: At twelve hundred and fifteen feet, they had not reached the *third* sandstone; and the general belief among practical men on the flats is, that none such is to be found there. But they passed through shale, slates, soap-stone, etc., as above the second layer. This well is, at least, four hundred feet deeper than any other whose record I was able to examine.

In passing, it may be mentioned here that a company has been formed at Petroleum Centre, with Mr. D. W. Davies as superintendent, to sink *a shaft*, about seven by seventeen feet, as far down as practicable. The organization is known as the Shaft Company. The experiment cannot fail to be highly valuable to the cause of science, and may repay all outlays upon it one hundred times over. Every company and land-owner in Petrolia ought to encourage the attempt. The evolution of gas will constitute the chief difficulty to the progress of this undertaking; and may prevent the introduction of artificial lights altogether. As, however, it is much lighter than common atmospheric air, it is believed the latter can be forced down in quantities sufficient to render the operations innocuous. Much interest will attach to the work as it progresses.

Mr. T. S. Truaire, a refiner and a gentleman of veracity and much intelligence, furnishes the subjoined statement from his record of a well, sunk under his direction, one mile above Oil City: "At the depth of two hundred and five feet, we struck the first sandstone, and went through it at two hundred and forty-three feet. From that point to three hundred and fifty feet, we found hard-pan clay, beyond which we struck the second rock, thirty feet thick. From three hundred and eighty to four hundred and fifty feet, we found another shale; and on passing this, we entered the third sand-rock, of forty feet in thickness. In this we struck a seventy-barrel well."

The points selected above are nearly eighteen miles apart, or fifteen in a direct line, the difference in elevation between their surfaces being nearly two hundred and fifty feet. It is reasonable to conclude that, while the underlying rocks are nearly horizontal, the layers in some places disappear altogether, and in others are modified, both as to quality and thickness. Several persons report that on the Watson flats, immediately above the oil-vein, they find a hard, flinty rock, a few feet in thickness, black and smooth, like that used for whetstones. Others report finding a mud-vein, from three to five feet thick, in the very heart of the second rock. It is supposed that this has settled as a sediment, where oil and gas formerly existed, after exhaustion of the former. Whatever the origin, it is productive of much trouble and loss in boring.

Mr. Fox, an experienced manager and a close observer, showed me numerous specimens of the inferior rocks. As compared with the first, a fragment of the third sand-rock at Petroleum Centre is harder, finer, and better polished, somewhat resembling a whetstone; its color is gray. Among the sand brought up from the bottom were found mixed black particles, as if of a carbonaceous origin. The specimen also had a distinct odor of the gas which comes off petroleum.

At Tideoute, on the Upper Alleghany, in sinking one hundred and fifty feet, they passed through layers of earth or gravel, slate, soap-stone, slate, gray sand, and white sand, finding oil in a gray, pebbly conglomerate. There and at West-Hickory, crevices were found, into which tools often dropped and got fast. On the Lower Alleghany, it is not customary to sink more than four hundred and fifty or five hundred feet, and the observations are of less value. One superintendent reports that below the second sandstone, they usually get shale and soap-stone; sometimes beds of hard shales; sometimes a third sand-rock is pierced, but not often. Another states that, on or near West-Sandy Creek, at the depth of five hundred and eighteen feet, they went through a substance of a bright, silvery appearance, which he believes to be a metal. It was very hard, and the bed about nine inches thick. The gentleman was evidently trustworthy, but he could only describe the substance from memory, having bored through it some years ago.

The wells put down five or six years since, in several cases got petroleum in the first sand-rock, their owners having been led to select such spots from "surface indications;" that is, petroleum oozing out of the ground. The famous Drake well struck oil at the depth of only sixty-nine and a half feet. In only some instances did men think of going down to the second rock, and in none to the third till long afterward. The article found in the uppermost layer is darker, thicker, heavier, and more valuable than that coming from the next, as the latter is apt to be superior to that obtained from the third rock. On French Creek, the splendid lubricating oil is got in the second series exclusively; the only instance in which I could hear of oil having been got lower down, showing that it was of the illuminating kind. But experiments, sufficient to prove the existence of a general law on this subject, have not yet been made; and of all regions, Petrolia is the last where a general principle can safely be inferred from particular facts, Nature having apparently taken a delight in setting all her own regulations at defiance. The only law which can be recognized with certainty is that of *lawlessness.*

If people wonder that sagacious and truthful men should sometimes vary so widely in their descriptions of the same object or phenomenon, let it be borne in mind that few of the superintendents, and none of the drillers, however experienced as operators, have received a regular scientific training; while professional *savans* have kept far too clear of the oil regions, as if they dreaded to come in contact with petroleum, except when their opinions were solicited and paid for by interested parties. This is not all the difficulty, however. The evidences of what the nether formations consist of, and what they contain, come up in the sand-pump, before passing into which the matter is ground fine, and being mixed with the water forms a paste or fluid. Seldom is a pebble bigger than a pea brought up by the ordinary process. Only when the workmen introduce some extraordinary agent like the torpedo, can they calculate upon getting fragments of the rocks to furnish *data* for observation and reflection.

For the same reason, the relation of petroleum to salt water is difficult to be gathered. The two liquids have been found, in almost immediate contact very generally, in first, second, and third sandstones; although it is now rare to get either in considerable quantities from the first. But the sand-pump coming up with a load of dissolved and mixed sand and clay, brings with these and the brine more or less fresh water which falls down from above, so that the precise spot where a salt-spring may be reached can not be ascertained. All that we know with tolerable certainty is that, on passing the second layer of sandstone, little by little the water begins to taste brackish; but not until the well has been "seed-bagged" and pumped for a time does it reach the full degree of strength, which is often equal to that of the strongest sea water. In some instances, this was struck, on Oil Creek, at depths of only seventy or one hundred feet. At Tideoute and West-Hickory, the salt was reached, in connection with the oil, at various depths between one hundred and ten and one hundred and fifty feet, the rule there being—no brine, no petroleum.

But why this close neighborhood on the part of liquids which do not mix mechanically, have no known chemical relationship, and are never found associated in either the animal or vegetable world, except as traces? It is a common proverb that oil and water cannot be made to mingle, yet Nature, in her subterranean laboratories, seems to delight in setting this rule at defiance; for while the brine usually manifests itself first in order, when the pump is applied, it never entirely forsakes the oil, the two clinging to each other like brother and sister. They are found together in West-Virginia, Ohio, and Kentucky, as well as in Western Pennsylvania.

And whence the brine? it may be asked. This question admits of a ready answer, and but one, by means of the fossiliferous remains and impressions of marine plants found in the sandstone, all going to show that every successive layer of it and of clay shales was formed in a shallow sea, which kept sinking gradually, as the coasts of certain countries are known to be doing at present. All this must have been going on through an untold succession of ages before the upheaval of that portion of the American continent began, or any river had begun to thread its devious course along the slight depressions. Added to the above is the following remarkable fact, stated by Mr. Ludovici, of the Humboldt Refinery, near Plumer, a most intelligent gentleman: From one of their wells, about seven hundred feet deep, was brought up a fibrous, yellowish substance, *closely resembling salt meadow-grass*, and not quite decomposed. The inference is clear that the sand-beds and clay deposits, the latter of which turned to shales, cracking and cleaving, as we may behold them, under the influence of heat, any day, were, in their "half-baked" condition, completely saturated with the salt water, which remained in their seams and crevices long after the great valley had been upheaved from the ocean-bed to its present elevation.

While on this subject, I may relate a curious experiment made by Mr. Morrell, who placed a quantity of petroleum and salt water in an atmospheric pump, and then exhausted the air, having previously shaken the two together, so as to mix them as perfectly as possible. The result was, that *the water settled to the top and bottom, the oil remaining in the middle of the vessel.* This is the more worthy of notice since it conforms to common experience in the oil regions, brine coming first up in the pumps, next petroleum, accompanied by gas, and lastly salt water. Why this disposition in the tube? Who can explain the cause of capillary attraction?

On the geographical and geological relations of coal and petroleum, I have already made some general observations. Even with the imperfect information within reach, the subject has a practical bearing. It is very certain that companies issuing prospectuses which intimate that *because* coal exists on their property, *therefore* oil will likely be reached, reckon without their hosts. The intervening space between the two (from four hundred to one thousand feet) is so great, and contains such an immense variety of rocks, that I regard the passage of petroleum downward by filtration as an impossibility. If the heavy sea water, for example, existing in the second series, has in thousands of years been unable to work its passage downward to the ocean-level, is it to be supposed that beds so completely filled with it would admit the entrance of a lighter liquid *from above?* Fresh water, we know, is scarcely ever found more than two hundred feet beneath the surface of the rivers, and never at the depth of four hundred feet, except by passing down the oil-wells. Could petroleum, still lighter, have made its way through slate, hard-pan, soap-stone, and all the other sedimentary formations, passing through a band of stone so hard and unfractured that it would seem to have been thrown like a lid upon the precious deposit, to keep it down? If it now bursts its barriers with such violence upward, would it enter them by the mere force of gravity? These are not the only difficulties in the way of that ready theory which traces the creation of oil to the distillation of coal. A still greater difficulty is to account for the existence, in such quantities, of inflammable gas. Did the carbonetted hydrogen—so many times lighter than atmospheric air—also work its passage downward? If it be replied that, in common with petroleum, it was extracted from the coal at such low depths by subterranean heat, applied subsequently to its deposition, I ask again: *Would not this heat have decomposed the water*, which would have parted with its oxygen, and thus have converted the gas into carbonic acid, which is poisonous and not inflammable, instead of carbonetted hydrogen, which has the very opposite qualities? No theory is hedged in with difficulties so numerous and insuperable as that which traces to the coal-fields the existence of rock-oil.

But if the manifest truth be admitted, that previous to the formation of our upper sedimentary rocks, an atmosphere, containing its present constituents of oxygen, nitrogen, carbon, and hydrogen, only in different proportions, perhaps, must have enveloped our globe, it is easy to see that "from the beginning," how far back soever we fix that date, hydro-carbons of various kinds must have been formed, whenever and wherever the temperature of the earth s surface permitted this. Whether coming into existence, as bogs, fens, forests, and the like, to be converted by pressure into coal-fields and thence distilled into petroleum and gas, or whether manufactured directly by heat, which might be generated by mere pressure or force as the equivalent of heat, in the laboratories of nature, we know nothing. One thing, however, seems probable, if not certain, namely, that as coal takes us back geologically to a carbonaceous era anterior to existing bog or forest, so petroleum discloses to us another such era equally anterior to coal, at least anterior to the coal found on this continent. And with this hasty attempt at solving what may, reasonably enough, be regarded as beyond the reach of an ordinary observer, I dismiss this part of the subject.

At the same time, I would most earnestly invoke men of science everywhere to give it a more thorough examination than it has hitherto received; to come to the oil regions, and remain there for weeks and months, collecting pebbles, fossils, fragments, and all other materials obtainable from the nether world. Let them spend their time and labor as enthusiastic explorers of truth, not with a view to lend their names to this or that Mammoth Gas Bubble Company, for a consideration in dollars or dollars' worth, thus fastening a stigma upon science, as indolent and behind the age, while it panders to deception, if not by misrepresenting some facts, at least by a studied concealment of others.

CHAPTER II.

APPEARANCE OF THE COUNTRY—THE CLIMATE—CHARACTERISTICS OF THE PEOPLE.

THE oil region of Pennsylvania is entered at four principal points, which may be termed the natural gateways of the country. Two of these the Alleghany River affords, it being navigable on the south from Pittsburgh, and (occasionally) on the north from Irvine, on the line of the Philadelphia and Erie Railroad. Rafts and flat-bottomed boats, indeed, come down from much higher points during spring and autumn. While a considerable proportion of the imports come by way of Pittsburgh, and large quantities of petroleum, both crude and refined, are daily sent down to that city, the great volume of travel to and fro passes by railroad. The Atlantic and Great Western Railway proceeds in a direction nearly parallel with Oil Creek, and at the average distance of about thirty miles, to the westward. The points affording communication with this trunk-line are Corry and Meadville, about forty miles apart, the former also touching the Philadelphia and Erie, now operated, under a lease, by the Pennsylvania Central Company.

Taking the cars of the Oil Creek Railroad at Corry, the passenger is apt to find himself inconveniently packed by the way, and may not, indeed, be able to procure admission further than the platform, feeling only too happy that he is not among the disappointed company who have been left behind. After traversing an upward grade for a few miles and the table-land beyond, he finds the road entering a branch of the famous Oil Creek. Passing near Oil Lake and the village stations of Centreville and Hydetown, he at length reaches Titusville, distant twenty-eight miles, the two hours' ride costing only one dollar. At the dépôt he may bid adieu to cheap fares, good beds, clean sheets, and other characteristics of civilization "in the States."

From Titusville the railroad proceeds down the valley to Shaffer's Station, nearly eight miles; but most of the passengers stop off at the former, it being the business centre of the upper end of Petrolia, and the point from which future operations of any kind can best be carried on.

The other gateway is Meadville, from which a branch railroad has been built by the Atlantic and Great Western Company to Franklin, twenty-seven miles, and thence to Oil City, seven miles further, the latter section under the charter of the Oil Creek Railroad Company. Between Meadville and Franklin this route follows the eastern bank of French Creek; and from Franklin to Oil City, the south-western bank of the Alleghany. Around Meadville, which is very pleasantly situated among the hills, and nearly all the way to Franklin, the country has been cleared and is under cultivation. The bottom-lands, though here and there washed away by recent floods, are as inviting in quality as the prices of farm produce are te[illegible]ing in amount. The uplands have also been much m[illegible] generally reclaimed along this line than on that betw[illegible] Corry and Titusville. The line of the Philadelphia and Erie, from Corry to Irvine, may be regarded as a medium between the other two. In general, it may be observed that the wide distribution of boulder-stones over the surface, and the difficulty of ascent and descent of the uplands, constitute more formidable impediments to farming than the cold or tenacious nature of the soil. On some of the uplands or slopes, thirty bushels of wheat, and more than twice that quantity of corn to the acre, have been raised.

At either Titusville or Oil City the stranger finds himself in a new world, this impression being no way lessened by hearing others speak about the latest news from "the States," or returning to them. This change addresses itself to every sense. The objects which he is too apt to touch, in spite of all precautions, have a greasy, clammy feel. His nostrils are assailed by gaseous odors, such as they probably never before inhaled in the open air. Into his ears is continually poured a stream of speech, in a dialect essentially different from that taught in Webster or Worcester. Such phrases as "surface indications," "dry territory," "developed territory," "oil-smeller," with the names of a dozen implements unknown to the outside world, all uttered with earnestness and volubility, at once set his half-bewildered wits at work in quest of their meaning. He tastes petroleum and salt water, of course, to satisfy his curiosity or acquire information of their qualities. Then he sees—what does he not see, in the line of novelties?—tall derricks and huge tanks standing on side-walks or in gardens; engines running and walking-beams moving sedately up and down in the midst of what remain of the original forests; drilling apparatus at work; immense flat-boats or rafts floating down-stream with the current, or drawn upward by three or four horses abreast, plunging along the bed of the creek or river. If the weather be cold, these poor creatures will be seen not only straining their muscles with desperation, as the inhuman driver applies the lash, but with their manes, tails, and sides thickly incrusted with *ice*, formed from the water splashed up, as they stumbled in the river-bed. If it be later in the season, he may behold a mile in length of boats rushing violently down-stream, that being the day when an artificial freshet has been made for this purpose by the opening of dams in the upper part of Oil Creek. As preliminary to all these novel spectacles, he has been treated to the filthy streets and wooden side-walks of Corry, Titusville, and Oil City, the last bearing away the palm in point of disarray and disgust. He has also been made acquainted with the luxuries of hotel life, especially in regard to sleeping accommodations, with from four to ten straw beds in a single room, each tenanted by one, two, or three sufferers, according to the pressure exercised by the travelling public. On the parlor floors he has learned to become reconciled to an inch deep of mud or dust, while leathery beef-steaks are no longer regarded with contempt; for with its many disadvantages, Petrolia has the one transcendent merit of creating a vigorous appetite.

With very little loss of time he takes to exploring the valley. I shall assume that he begins with the region back of Titusville, that Pennsylvania Venice, arising out of the mud, which in April is still sufficiently deep and liquid to float a whole navy of gondolas. If the side-walks are a little uneven, let them not be despised; for the time is coming when a single plank will elicit an out-burst of welcome, as a god-send. Then there is Oil City, at the mere mention of which, Titusville is transformed into a capital with all the charms of Dublin or the neatness of Philadelphia. In the heart of that borough he finds large hotels or caravansaries by the half-dozen, and as many more in course of erection. On every hand new houses are rising under the incessant blows of the carpenter's hammer. At morning, mid-day, and evening, the screams of steam-whistles at the various machine-shops, foundries, and refineries, are painfully long and loud. In the various houses or sheds thrown up on the principal streets, where lots sell at New-York City prices, he finds whole platoons of land agents, lawyers, speculators, the agents of merchants and manufacturers, whose wares are likely to be in demand there, "drummers" of all kinds and from all parts of the country. He can hardly turn a corner without being "drilled to the third rock" by a pair of keen, inquisitive eyes, followed by the inquiry: "Do you wish some first-rate oil territory, sir?" "I would like to sell you a fourth interest in a fifty-barrel well." "Can't I furnish you with Jones's new patent blower or an Excelsior steam-engine?" In the midst of such interrogatories, it is gratifying to learn that Mr. Smith yesterday "struck" a two hundred barrel well on Cherry Run, and that Mr. Brown's has doubled its yield since he had it "reamed" out and that new "blower" put in.

Determined, however, on piercing the heart of the country, he hires a horse at ten dollars per day, and sets out on his pilgrimage down the valley. Immediately below Titusville, and above the confluence of the east and west branches of Oil Creek, he enters the celebrated Watson flats, a short distance beyond which he observes the derrick of Colonel Drake, erected in 1859, the first work of the kind in Petrolia. More than one hundred others, new and old, may now be counted within one mile of Titusville, especially near the point of confluence. Every thing betokens disorder, disarray, indifference to all except the one grand object of pursuit. There are no roads, no fences, and scarcely laws or regulations, except a few laid down in the leases or imposed by common consent. In a place where seemingly *meum* and *tuum* are confounded; where every man appears to act for the day, regardless of the morrow, one might reasonably suppose that violence and even bloodshed would be matters of almost hourly occurrence. So far from this, however, I am happy to say that there is probably no place in Christendom where human life is safer, and less danger to "portable property" exists, except from freshets and extravagant charges, than in and around Titusville.

Proceeding down the creek, where one's best road, off the railway, is the uneven bottom of that impetuous stream, the valley is found to grow quite narrow—barely one hundred yards from bluff to bluff. On the heights, overhanging the railroad and creek, where heavy forests of pine, hemlock, or white oak once grew, little now, save brushwood and stumps with long, horizontal lines of shale and sandstone behind are visible, the scene being here and there diversified by a small unpainted cabin, or by the ubiquitous derrick. Between the precipitous heights the creek describes numerous sinuosities, always apparently butting its head with full force against the steepest banks; in reality having made them such by dashing with such impetuosity against the rocks underneath.

Here, as well as higher up, one meets tall gentlemen,

encased in tall, shining boots, or what were such in their primitive state; wearing tall, black coats, tall, black beards, and carrying tall, black valises. They are adventurers, in search of lands, appointments, interests in wells, or individuals, whom they can sell and deliver equally with their property.

For the five or six miles immediately below the Watson flats, little boring has been done—ten or a dozen wells to the mile or so. The section is pronounced "dry territory." At Miller's farm or station symptoms of more activity are manifest; and at Shaffer's, where the railroad terminates, a cluster of hotels and another of shipping-offices have sprung up. To that point boats, filled with petroleum, in bulk or in barrels, are dragged up-stream at nearly all seasons. The arrangements for unloading this from the boats, hoisting it (by means of horse-power) to the top of a high trestle-work, and thence conveying it to the extensive sheds which adjoin the railroad, are extensive and well adapted to the purpose. The only drawback to the whole is the cruel treatment of the horse, whose lot has been made worse by the great discovery, though its benefits have been felt by whales disporting themselves in the Arctic seas. One's first impulse is to curse the day petroleum was first discovered, and to knock down the barbarians by whom the task of applying the lash has been voluntarily accepted.

Beyond Shaffer's the nearly level bottoms begin to widen, affording the creek more abundant space for its frequent gyrations. Each of the half-moon flats beyond has a distinct name, usually given it after that of the former proprietor, in connection with the farm lying immediately above and behind. Every flat has also its system of derricks, and, in general, characteristics of its own distinguishing it from those above and below. In passing downward the derricks will be seen to hug the bluffs more closely, and even to climb them, in places, to the height of one hundred or one hundred and fifty feet. For about midway between Titusville and Oil City the stranger has entered the great heart of the oil region, where the Sherman, the Noble, the Empire, the Craft, the Wild-cat, the Jersey, the Coquette, and other famous wells were formerly, or are now, wont to discharge (by flowing) their hundreds of barrels per day. Each of these famous producers has its own street or block of black, dirty, greasy tanks, from two or three to ten or twelve in number, with an aggregate capacity of between five thousand and fifteen thousand barrels. Most of these are roofed over and located close to the creek, with a view to easy loading in the barges. From their bottoms exude streams of the dark green liquid, which crawls along by slimy paths to the creek, covering its entire surface with a film of petroleum. Many barrels of it thus escape every day, to the deep regret of the looker-on, who wishes he had the facilities, with the right to use them, for preventing such a waste. With a clear sky overhead, the different hues formed by this "oil cast upon the troubled waters" are exceedingly delicate and beautiful, and can hardly have failed to suggest the extraction of certain rich and rare colors, as analine, from this wonderful product.

On every farm henceforward is a village, bearing the farm name, or the affix "ville" as a substitute. The names and the settlements are, indeed, about equally outlandish. If the one be prosy, the other is slatternly, in muddy weather indescribably so. Thus, within the space of ten miles, we have the names of "Funkville," "McClintockville," "Tarr Farm," "Rouseville," "Rhind Farm," as well as other places of less repute. "Petroleum Centre," as it has a decent name, has a pretty situation, a good bridge, and is laid out with some degree of regularity. But as its site is the prettiest, so the naming of it would seem to have absorbed all the poetry in the valley. The village usually consists of a number of oil companies' offices, about twice as many boarding-houses, perhaps a school-house, where religious services are held occasionally on Sundays, a hotel or two with their wonted accommodations, in the shape of sitting-rooms and bed-rooms, for which the modest price of three to four dollars per day is charged. In Rouseville, Plumer, and one or two other points, banks have been established. Post-offices abound, nearly every farm having one; and the telegraph extends to every nook and corner in the country as fast as a good well is struck. The number of houses in these villages or hamlets ranges from ten to fifty, and the population, exclusive of strangers and pilgrims, from one hundred to eight hundred. The houses are built of weather-boards and strips only, being guiltless of paint on the outside or of lath and plaster within. On some farms the streets are laid out with a fair degree of order, and the more elevated spots are selected; in others, the law of lawlessness prevails, and a goodly number got immersed in the late freshet. As to drainage, fencing, shrubbery, or gardening, these are all in the future tense and conditional mood. Once, and only once, I did notice a discharged soldier engaged in planting a little grass-plot in front of his cabin. Probably one half the engineers and laborers sleep in cribs attached to the engine-houses, and some even cook their own meals there, in order to escape a charge of seven or eight dollars per week for board and the coarse accommodations had in the boarding-houses. If there is one cow in that part of Petrolia, she escaped my observation. Even the dog-tribe are far from being numerous.

Nearly every mile along the lower part of the valley, a "run" discharges its waters into the creek, running at the bottom of a ravine, more or less deep and wide, according to the volume of its waters. The principal of these entering from the east are Bull Run and Cherry Run, (the latter at Rouseville, three miles above Oil City.) On the west side are Bennehoof, Cherrytree, and Cornplanter Runs. On Cherry Run, however, are more works than on all the other tributaries of the creek; while in point of productiveness it disputes with Oil Creek the claim of supremacy. At its mouth, and for some miles above, the derricks stand as thickly as the masts of shipping in the East River, at New-York. They evidently mean to dispute possession of the uplands with the squatter sovereigns.

One of the most remarkable phenomena of the valley is the large proportion of idle wells—in some localities at least nine tenths, and in the most active three fourths. The visitor will hear various reasons assigned for this unexpected idleness—reasons which I propose to discuss elsewhere, contenting myself for the present with noticing the fact.

The highway is coëxtensive with the bottom; and if that be not found sufficiently capacious, the traveller is at full liberty to annex a portion of the bluff or table-land. As the soil is clayey, and as it rains every other day, during the spring months, while literally *no* attention is paid to the roads, (so-called,) the reader will please fill up the picture, as to travelling facilities, to suit his own taste. If to the sticky paste be added wriggling rivulets of water, coated with grease, settling in numerous basins or puddles, across which one has to work his passage, by leaping from prostrated tree-stem to stump, plank, rail, iron tube, stone, old boiler, walking-beam, or whatever other object he can reach—these operations being accompanied by an occasional downfall, and a frequent splash of mud and oil up to the hat—perhaps the indescribable enjoyments of a foot-march through Petrolia may be conceived. But I had almost forgotten that the stranger who has been thus far conducted through the country is supposed to be a gay cavalier, not "doing" the region on his own nether extremities.

As if to afford frequent opportunities to wash and be clean, the creek cuts across the highway, that is, the valley, every half-mile or mile. The horse can ford it without trouble; but for the humble pedestrian there is no means of crossing save by a ferry-boat or rope-ferry, where the toll is five cents for each trip. These "institutions" are chartered under authority of the State, to whose treasury each pays an annual tax of ten dollars. For this paltry sum he secures a monopoly for a long distance above and below; and as it sometimes happens that, after a good well has been struck, a new ferry must be started, it becomes necessary to pay Charon No. 1 a "royalty" of so much on every passenger transported a distance of seventy-five feet! I fully agree with the observation made by one of these boatmen, that he would not exchange his skiff for a good oil-well.

Added to the natural disorder prevalent in the oil regions, are the wrecks produced by the great freshet of this spring, the most destructive that ever visited any portion of the Northern States. It not only swept down the river numbers of houses and immense quantities of petroleum, but deposited all along the low lands fragments of boats, dwellings, engine-houses, furniture, fuel; overturning derricks, carrying off wooden platforms laden with engines, and hurling the whole with resistless force against bridges, which shared the common fate. The side-walks in Oil City have been left wherever the capricious element chose to deposit them; a huge flat-bottomed boat was dropped in the principal street; dwellings and factories were lifted from their foundations, and moved hither or thither—perhaps to encroach on the thoroughfares, perhaps to stand at a different angle to them. Within half a mile of the built-up portion of that "city," boasting of its burgess and council, the carcasses of no fewer than twenty horses, which had perished during the flood, were suffered to lie unburied nearly a full month, and may be perfuming the atmosphere to this very day!

Oil refineries, belching forth clouds of black smoke, or (as is quite common) lying idle, form one of the features in the landscape of those valleys. They are for the most part small establishments, each with a capacity not exceeding three hundred barrels per week. The mode of treating petroleum, so as to prepare it for use, is explained in another chapter.

The stranger will quickly master the difference between wells in progress and those completed. The engine-houses and derricks of the one are comparatively clean and white; after striking oil, however, they get coated with the universal pigment, and turn as black as the smoke rolling out of the chimneys. Men's clothes are, of course, in keeping with the general scene. Waterproofs are made without cost in Petrolia.

The various tributaries of Oil Creek, and the Upper Alleghany, are too small and rapid to afford facilities for boating; hence, the oil has to be conveyed by wagons, which are hauled over roads, whose equals exist nowhere else. Formed originally by the teamsters to suit their own convenience, they are kept in order only by the *débris* which is washed down from the heights, and remain floating masses of slush, huge fragments of rock filling more or less of their unfathomable depths. It would appear that the only conceivable way of mending their ways in Petrolia was to apply fresh curses and kicks to the poor horses.

Arriving at that perfection of filth and disorder, Oil City, the visitor finds a newly extemporized borough, a mile and a half in length, which hugs the base of the heights west of Oil Creek; then crosses it, and pushes up the face of the eastern slope for one hundred and fifty feet, where at length it begins to expand itself, as if conscious that it can do so safely for the first time. The view presented from "Cottage Hill" is certainly picturesque, and the contrast presented between that charming spot and the slovenly avenue beneath cannot fail to make a profound impression. The noble Alleghany, three hundred yards wide, sweeping along the bases of the hills, and receiving, not only its tributes of numerous creeks, with their many-colored waters, but scores of barges, steamboats, and other vessels, constitutes a most attractive part of the scene, the effect perhaps heightened by the physical lawlessness of the lower city. In the distance round-shouldered and pyramidal hills, their angles and terraces standing out sharply, constitute a grand background to the picture.

Up and down the Alleghany for about twenty miles; along French and Sugar Creeks for half that distance, as also along the various tributaries of all those waters, within the distance stated, the scene is similar to Oil Creek, though mostly in a less marked degree.

THE CHIEF TOWNS.

The principal centres of population and business in Petrolia are the following:

Corry, situated at the point where the Atlantic and Great Western Railroad intersects the Philadelphia and Erie, and where the Oil Creek line terminates, is a thriving village, containing between three thousand and four thousand inhabitants. Half a dozen years ago it had neither a local habitation nor a name, and scarcely as many log-huts in the heavy forests. Indeed, it is still literally "in the woods," the valuable portions of which have alone disappeared, leaving the stumps and roots standing in the principal thoroughfares. Corry is pleasantly situated, is regularly laid out, and is fast becoming a prominent business point. The establishment of the Downer oil-refinery there, (the largest in that region,) by some sagacious men from Boston, gave the place a great impetus, which promises to continue for some time. Ordinary sized lots on the principal street sell at from two thousand to three thousand dollars each. In no part of the country are more new houses in progress. Corry contains two banks, four churches, one respectable hotel, public schools, a lecture-room, (in which poor concerts draw better than lectures,) a newspaper-office, from which a weekly independent journal (*The Telegraph*) is issued. In its population the Eastern element predominates.

Titusville, named after one of the early settlers in that valley, contained about one hundred and fifty inhabitants before the oil excitement; it now contains probably five thousand, besides a considerable floating element. It has four or five refineries, barrel-factories, machine-shops, and foundries, (the whole employing nearly five hundred men;) also two banks, the usual assortment of churches, (which appear to well sustained,) a theatre, and a large number of hotels. The only newspaper issued is a weekly, (*The Reporter*,) which is independent in politics, and conducted with spirit. The proprietors contemplate the establishment of a daily shortly. The situation of Titusville is too level and low to be easily drained; and in fact the attempt would seem to have been given up. The place is regularly laid out, however, and the outskirts are decidedly attractive. Population orderly, enterprising, and largely on the increase. The derrick is already beginning to make inroads on the gardens. If successful in the search after oil, there is no telling to what figure the price of lots will advance. Already they are higher than in Corry.

As *Oil City* has the most disgusting *name* in all Petrolia, so every thing else is in keeping therewith. One of its first and best sustained institutions was a race-course, laid out on the summit of the highest hill, "for improving the breed of horses;" while those wretched quadrupeds were left to flounder, lie down, and die on the horrid thoroughfares, termed streets, below. Folly kills itself;

for, except on Cottage Hill, the place has almost ceased to grow. Even with the extension of the railroad thither, it is "a finished town." Population, about five thousand, besides a thousand or two of floating elements. Oil City has two banks, half a dozen hotels, ten oil refineries, four or five churches, (including those in progress,) and a public school building, nearly completed. Two weekly newspapers, representing the great political parties, are printed there. The creek at Oil City is about sixty yards wide, and is crossed by a trembling structure, termed a bridge, which the authorities permitted a company to erect on the foundations of the one swept away, charging five cents for the privilege of crossing! The creek divides Oil City into two nearly equally large sections; though that on its western side is more populous. A suburb is rising on the opposite bank of the Alleghany, which is connected with "the city" by two rope-ferries. The place is altogether a new creation. An interesting fact in connection with it is, that an Indian tribe has recently put in a claim for the ownership of two thirds of its site, which, it is asserted, was granted by the State to "Cornplanter," a noted chief of the Senecas, about thirty years ago. The matter is about to be litigated.

Franklin is situated at the mouth of French Creek, seven miles below Oil City, and contains nearly three thousand inhabitants, besides a large ingredient of "drift." Its site is on the whole pleasant, and the streets are spacious and regularly laid out. Having been a county-seat before the era of petroleum, its public buildings (court-house, churches, etc.) are more substantial than either stylish or outlandish. New churches and public schools, both spacious and elegant, are among the improvements contemplated at an early day. At different points in the village derricks have arisen; but many of them are now idle, and withal far from being attractive objects. Franklin is situated on the south-west side of French Creek, which is there one hundred yards broad. The lower bridge crossing it was carried away by the flood; but the magnificent suspension bridge across the Alleghany escaped. Franklin has two weekly newspapers, which appear to be well supported. The bulk of its population still belong to the old stock of settlers. Excepting a barrel-factory, and some small oil refineries, (on the other side of the creek,) the manufactures of the place are of little account. As a distributing point, it ranks next in importance to the places already noticed.

Franklin boasts of an antiquity of a full century. The point below the confluence of the river and creek was selected by the French for the site of one of their chain of forts connecting their Canadian possessions with Louisiana; but of that work not a trace is now visible. In 1754, General Montcalm visited the place, and in his report of it took occasion to describe the war-dances and religious worship of the aborigines. Among other matters he refers to their mixing oil, gathered from the neighboring creek, with their war-paint; also to their use of it in sacrifices, kindling it with torches, at the sight of which they set up a shout that made the valleys ring. Strange that a whole century should have elapsed before the paleface set up *his* shout over the discovery.

Meadville is the largest, prettiest, and most cultivated place in the oil regions, having between eight thousand and nine thousand inhabitants. Its situation on a tributary of French Creek, surrounded by gently swelling and well-cultivated hills, is equally healthy and attractive. Meadville is noted as a seat of learning, having, besides an excellent system of public schools, two colleges, one (Alleghany) belonging to the Methodist, and the other (a theological school) to the Unitarian denomination. The opening of the Atlantic and Great Western Railway has contributed greatly to the progress of that place; and it derives a portion of its prosperity from Petrolia, though twenty miles west of the field of operations. Three weekly newspapers are printed there, and the Railway Hotel claims to be the best in the world.

Warren, like Corry and Meadville, lies on the outskirts of the oil region, being about fifty miles above Oil City by river. Navigation, however, is open only to Irvine, and that after a rise of water. Warren is an old county town, on the line of the Philadelphia and Erie Railroad, and contains nearly twenty-five hundred inhabitants, with the usual public buildings and newspapers. It has just begun to feel sensibly the ground-swell of the petroleum development. A number of wells are in progress near Warren, and, if successful, it will leap upward like Corry or Titusville. *Irvine* is an unpretending railroad station, seven miles below Warren, with a well-kept hotel, and two or three small factories.

Tideoute is an old village, or rather two villages joined together, pleasantly situated on the west bank of the Alleghany, fourteen miles below Irvine. It has two churches, hotels, barrel-factories, and about two thousand inhabitants. The famous Economy wells lie on the opposite side of the river.

Tionesta is a growing village, with perhaps five hundred inhabitants, about ten miles below Tideoute. It is situated on the east side of the river, at the mouth of Tionesta Creek. East and West-Hickory are incipient villages higher up-stream, as is President lower down on the Alleghany. All these have their groups of wells completed or going down; and may be said to have been called into existence by the discovery of petroleum.

THE CLIMATE

of that section of Pennsylvania is a subject of universal complaint in the spring season, when it is certain, the inhabitants say, to rain at least every other day. There is a good deal of truth in this remark. From the middle of March till the latter part of April, there were never three consecutive days, and seldom two, without rain or snow falling. As throughout the whole Mississippi valley, the prevailing winds are from the cardinal points, the south and east gales being hot and moist—in other words, bringing thither from the Atlantic and the Mexican Gulf the moisture-laden vapors, which, upon being struck by dry, cold, currents from the north and west, give out their superabundant moisture in the form of rain, hail, snow, etc., according to the season. The atmospheric current preceding such a visitation is invariably succeeded by one blowing from the opposite direction, except where it is deflected from its course by mountains or waters. In fact, it would appear as if the precipitation of rain were directly due to two currents blowing in contrary directions at the same time—the lower from the south or east, and the higher from the north or west, and that with the gradual descent of the former toward the earth's surface the wringing-out process in the latter took place, thus creating a vacuum, into which a fresh gale rushes from the ocean.

The dry, magnetic winds from the interior would soon lick up the moisture not carried off by the streams, were it not for puddling of the surface to so great a depth by animals and vehicles. These general observations extend to the entire heart of the continent, as well as to the oil region of Pennsylvania.

In the latter, however, a local influence exists, such as is not so sensibly felt further westward and southward. At the lower end of Lake Erie, large quantities of ice remain long after the Western rivers, and even the upper portion of that lake, have been cleared. The hot, moist winds from the Atlantic naturally make for the lower end of the lake, where they assist in thawing the ice; but in so doing are themselves wrung out by coming in contact with the belt of cold atmosphere immediately above it. Hence those frequent changes in the direction of the winds at that season, the numerous rainy days or parts of days, not making, in the aggregate, perhaps more inches of water than elsewhere under the same latitude; but coming at short notice, and upsetting calculations, make living there more unpleasant than it otherwise would be. One interest alone, the hotel-keepers, reap large profits from this dispensation; while the servants are oppressed with double duty if they attempt to preserve cleanliness in spite of muddy boots, muddy clothing, muddy luggage. If it be true that every man must eat his peck of dirt some time, in no other region is there an opportunity for dispatching this task so hastily as in Petrolia.

THE PEOPLE.

That man should be superior to his accidents is, we are told, the fundamental principle of true democracy. On the whole, the Petrolians are superior to their surroundings or circumstances. It takes time, however, to study their good and less good points of character. The stranger who visits that country in quest of fortune or information, is apt to form erroneous conclusions respecting them, the first few days. At the principal gateways leading thither, he is certain to encounter a class different from the great body of operators in the valleys. The former consist largely of hangers-on about hotels and boarding-houses, who are in quest of victims; of roystering, blasphemous teamsters and boatmen; of disappointed fortune-hunters, preparing to return home, and having a very low estimate of life and manners in Oildom. Everywhere he finds exorbitant charges, without an apparent disposition to oblige. If he be a religious man, he will be hourly shocked by profanity; if a humane man, at the brutality with which the lower animals are treated; if a man of generous instincts, at the intense selfishness, the sordid love of gain, so widely prevalent; if a man of taste and culture, at the outlandish condition of the houses and the streets, with the indifference of the people toward intellectual pursuits, beyond the immediately practical. If he proposes to introduce any other topic of conversation beyond the never-ending, still-beginning themes of oil and war, oil and politics, he will presently find his company thinning out. Ten minutes' loud conversation on philosophy, literature, science, or religion would give him full command of a parlor, or even a bar-room. For the inhabitants of those large towns removed thither to make money, and do not mean to be turned aside from the one grand object of existence.

It would be wrong, however, to judge the entire population by this ingredient. Nay, (in Titusville,) I met *one* saloon-keeper who keeps up his old reading habits. Having pierced the crust of mere adventurers, speculators, and peculators, bespattered men and dowdy women, let the visitor traverse "the rural districts," and he will discover intelligence, refinement, even generosity. *As a class*, the superintendents of the large companies are gentlemen of culture, who would adorn any society. Not a few of them were commissioned officers in their country's service, who have gained honorable distinction ; some were conductors of newspapers, who have carried with them to that solitude their abundant knowledge of men and measures appertaining to the outside world. They are ready, at all reasonable times, to impart useful information, and the observations gathered by this body of officials ought to be regarded as treasures of no ordinary value.

And in point of kindness of heart and readiness to oblige, the engineers, drillers, and others engaged about the works will compare favorably with any other body of men I have ever seen. Where they could not give the trustworthy information sought, they were ever ready to put me on the trail after it. Of hoggishness, or a deliberate purpose to deceive, not one in fifty could be justly charged. Can the outside world produce a cleaner record?

Not that any class of employés are perfect in every respect. The officers are supposed to comprehend clearly more than one mode of raising *gas* as well as oil ; and in many instances, I fear, are too ready to wink at the bad schemes of stock-jobbers and speculators, if not to lend them active support. A kind of lax morality prevails that misrepresentation is, if not justifiable, at least excusable, when committed in the interest of one's employers. Hence the vast exaggeration in the yield of wells, and the studied concealment of facts that would injure the sale of stocks or interests at fictitious prices. The Spartan law, as a social regulation, is still too generally obeyed; for most of us chuckle when we hear of a dishonest operation, provided the performer has been *smart* and successful in his stratagems to pick other persons' pockets.

No community on the face of the earth has a smaller proportion of drones to the number of working bees than Petrolia. This observation applies to city, village, and single shanty. Nobody but has a hand engaged in *some* business or pursuit; many in half a dozen. If a man betakes himself to mercantile life, he reckons upon giving it from twelve to fifteen hours per day, filling up his leisure moments with speculation or an agency. The young fellow who would stand at the street-corners elsewhere, there kills two birds with one stone by offering to sell wells, or interests in wells, or leases, or refusals to those whom he can button-hole. If Satan found mischief *only* for the idle, his occupation would be gone in the oil region. Perhaps the high cost of living has impelled the slothful as well as the diligent to this remarkable activity, but it seems to be an admitted principle on all hands that people have gone thither to *work*. On this account, the country is essentially orderly. Property as well as life is more secure than in any Eastern city. Even drunkenness is by no means as common as might be expected, in view of the rough-and-tumble modes of life prevalent. I have seen less of it in Oil City or Titusville than in country towns of the same size elsewhere. Yet I do not believe that one man in fifty is a member of the temperance association. It is said that the vice of drinking prevails to a considerable extent on Sundays on some of the farms, but the wonder is that it should not have become universal.

Last fall, intense excitement prevailed near Oil City, caused by the dead body of a resident having been found close by it. He had been murdered in open day. An indignation meeting held appointed a vigilance committee, and the whole population joining in chase of the criminals, they hastily decamped. It is supposed they were part of a gang from the East, who expected to "operate" in Petrolia as some of them had done in California.

During the first three or four years of the oil excitement, little respect was shown to the first day of the week, and few attempts at establishing Christian worship were made outside the focal points. The flowing wells poured out their wealth on that day as on the remaining six, and the pumps copied, as far as possible, after the others' example; so the people pumped, and barrelled, and drove, and shipped petroleum on Sunday as well as Saturday. Man lives not by oil alone, however, any more than by bread. A change has been gradually taking place in this respect, giving man and beast the advantage of a septennial day of rest. Perhaps this improvement was brought about by the men refusing to work; perhaps as a stroke of policy, to retain the more sober and steady portion of the mechanics and laborers; perhaps from conscientious motives on the part of the large companies. Sunday work seldom takes places now, except in wells which have been flooded, or are in danger of becoming unserviceable for a time, in consequence of the water getting the upper hand.

At Rouseville, Petroleum Centre, and a few other points, small buildings have been erected, to serve for chapels on Sunday and school-houses during the week. I am inclined to believe these are better sustained by commutations in money than by personal devotion, and that the worshippers are more disposed to purchase tickets to the Celestial City for their friends and relatives than to get aboard of the cars and ride themselves. Like some other matters of importance, this is left till the fortune-hunters return to "the States." The fact is, in Petrolia, the church universally believed in is an engine-house, with a derrick for its tower, a well for its Bible, and a two-inch tube for its preacher, with mouth rotund, "bringing forth things new and old," in the shape of two hundred barrels per day of crude oil, mingled with salt water. In the principal business-centres, regular societies have been instituted; but that practical Christianity which leads men not only to love and fear God, but love mercy and hate covetousness, is not in a flourishing condition. Indeed, I fear some of the "under shepherds" are more intent on oil development than in rebuking the vices and follies of the community; otherwise, it seems to me, profanity would be a little less common; some sympathy would be shown to the brute creation; selfishness and swindling would at least feel ashamed of themselves. I heard of a promising young divine who was making a good impression among his auditors, one of whom made him a present of a one-sixteenth interest in a well then going down. Oil was struck, and the gift was converted into twenty thousand dollars; whereupon the preacher retired on a competency. Let us hope that others will not thus be drawn aside by a glance at the hill Lucre.

The former proprietors in that part of Pennsylvania were largely descended from the Protestant part of the Irish population; and to this day retain many of the characteristics of their ancestors. As a rule, they are slow, steady, cautious, thrifty, and strong-willed. Nearly all have, on selling out, removed to Ohio or Western New-York, purchasing farms, and investing their surplus means in public securities. Many of them expect, after this whirlwind blows over, to regain possession of their farms at a tithe of what they pocketed from Eastern agents. The new-comers are a mixture from all parts of the country. California, Iowa, Kansas, Minnesota, and Missouri being represented with New-England, New-York, and Pennsylvania. Even Virginia, the Carolinas, Kentucky, and Tennessee have representative men there.

The Petrolians are nothing if not geological. Nearly every operator is ready to discourse learnedly on rocks, formations, strata, (in the singular number!) shales, sandstones, (comprising every thing from limestone to conglomerate.) As in nature, so in human nature—no two agree. A, after describing "a most remarkable phenomena," is positive that the best wells are to be found on the east side of all runs and "criks." B asks you to examine "that strata," and concludes that prudent men should bore only on the *slopes*. C, an old gentleman, fussy and seedy-looking, avers that the country is of volcanic origin, and is ready to point out certain rents in the hill-tops, through which Vulcan and his helpers found passages for the smoke and cinders of their forges; whence the petroleum. It would be uncharitable to surmise that either of these *savans* had a personal object in view, in the sale or leasing of land; yet stranger things have happened in Petrolia. But of all original characters, the most amusing is the ancient ploughman, wood-chopper, or flat-boatman, metamorphosed into a millionaire and a scholar. "This is

Sugar Crik," observed one of these newly extemporized linguists, "and that is a contributory to Sugar Crik; and the symptoms of ile is very premonitory!" Others are in the habit of pointing out the great benefits certain to flow from the further "envelopment" of the country. It is clear that Dame Partington has given some valuable lessons in Western Pennsylvania; indeed, who knows but that her ladyship has "oil on the brain"?

Health is the rule, and sickness the exception, there, in spite of the many drawbacks. Few persons exhibit the lean forms and sallow complexions so common in other parts of the country. On the contrary, as the men have a look of boldness and vigorous purpose, so they present the appearance of physical robustness in an unusual degree. This may be traceable, in part, to the circumstance that the hardships and privations felt there *drive away* the more feeble in mind, body, or purpose, who are thus strained out of the community. But there is more than this. The rough, wholesome, open-air exercise connected with this new life, the fresh mountain air, the fresh water pouring forth from a thousand springs, have built up the physique of hundreds of young men who previously languished behind desks and counters in the cities; have given them buoyancy of spirit as well as strength of limb, such as they never before enjoyed. The diseases to which strangers are said to be liable are principally connected with the digestive system, as diarrhœa, dysentery, etc.; but it is questionable whether these are traceable to the water so much as to exposure and over-exertion, especially working up to the knees in water, and remaining in damp clothes.

Of preparations for farming or gardening operations,

this spring, there are none. Speculation has become so rife that it extends to the uplands, which are accounted "too valuable" (such is the slang) for agricultural purposes. The little supply of milk that reaches the valleys, and nearly all the vegetables, equally with the supplies of meat and grain, come from great distances. The author believes that the best paying wells this year may be struck within eighteen inches of the surface, by drilling with a plough, reaming with a hoe, tubing with garden-seeds, and pumping with manure.

"*Olei sacra fames!*" The insane desire of oil *is* demoralizing. It leads to every imaginable kind of misrepresentation and cheating. In every transaction involving profit and loss, falsehood is expected, is looked upon as the rule, truth as the exception. This indifference to veracity and honor does not merely extend to matters connected with the oil-wells, but to those of every-day life—to engagements entered into by landlord and tenant, by mechanics, laborers, etc., whenever a slight advantage may arise by violating them. This "covenant-breaking," where no other obligation than a man's word exists, forms a topic of general complaint in Petrolia; and at this moment, it is not too much to say that no one expects his neighbor to certainly fulfil the conditions of a merely verbal contract.

CHAPTER III.

LOCATING AND SINKING THE WELLS.

In the earlier days of well-sinking, the inexperienced operator planted his derrick and drilled his well wherever he detected "surface indications" of petroleum, probably little thinking that it might show itself on the ground at a point far from vertical to its proper source in the sand-rocks. In general, the margins of rivers and creeks were preferred to spots more distant, even though equally low; hence the first crop of derricks grew up close to Oil Creek and the Alleghany. Even at this late day there is little to guide the adventurous operator beyond the conceded existence of *oil-veins* in the inferior rocks, which circumstance, however, could only become known by making numerous experiments. A new profession of men, claiming to be gifted with extraordinary powers, has arisen in Petrolia, namely, "oil-smellers" or "diviners." Let not the pious reader start with alarm, lest the practice of divination, (whatever it may have been,) condemned so repeatedly in the Mosaic code, has been revived in Western Pennsylvania and Virginia. No devil, demon, ghost, ghoul, fairy, goblin, or table-tapping spirit is known or believed to be at work, albeit the use of a twig of witch-hazel or peach might readily enough suggest to some the calling up of spirits from their vasty deep by modern enchanters. The mode of operating is substantially as follows: The diviner cuts from one of the trees mentioned, a bifurcated bough or twig, reducing the stem and the forks to about a foot in length, for convenience' sake. In each hand he grasps firmly one end of the fork, letting the stem point *upward* and a little inward. The hands should be held with their backs downward. With this simple apparatus off goes the "smeller;" and, on arriving above an oil-vein, it is claimed that the twig will turn round in his hands, in spite of his utmost exertions, until the stem points directly *downward.* It may be grasped so tightly that the rind will peel off by the operation; yet this will not prevent the revolution in his hands. The author once witnessed this operation going on in the hands of a gentleman of much intelligence and the utmost veracity, who was not a believer in the oil-smeller's claims or pretensions, yet had to acknowledge the existence of the phenomenon for which he could not account. It appears that the twig has not this remarkable power in the hands of all persons; for the author was unable to perceive any change or tendency in the wand in his own hand, on arriving at the same spot. Whether the difference were owing to magnetic influence or other cause, is unknown; as also whether the motion betokens the presence of water, petroleum, both, or neither. In England, it is said, the witch-hazel has long been used in this manner for the discovery of *coal;* in some parts of the Eastern States people try it to alight upon water-veins. In the oil-regions, some of the most productive wells have been located by oil-smellers; in more cases, however, their vaticinations of first-class works turn out mere moonshine. However, the diviners have become a power in Petrolia, among a people as keenly inquisitive and practical as are to be found, who reason in this way: "If there be any thing in oil-smelling, we may as well avail ourselves of it as not; for the diviner charges only from twenty-five to one hundred dollars for his services in examining a tract; and this is an inconsiderable item in the general expense, seeing we mean to bore any how." There is a pretty general impression that he is a better guide negatively than positively; that while oil may not be struck just where he directs, it is useless to sink where he has pronounced none to exist. In a word, the charmer, magnetizer, or natural magician has more real power among the operators than the latter are willing to openly concede.

The principal matters now attended to in locating a well are the following: The ground ought to be low, to make as little drilling as possible suffice; yet not so low as to be subject to floods. The lesson taught by the late freshet has been a most costly one in this respect. There should be sufficient space nearly level, for the derrick, engine-house, tank, etc.; this secured, they may be placed on the top of a knoll or the face of a bluff. There should be no hard boulders on or immediately below the spot where it is proposed to drill, as the driving-pipe must descend perpendicularly. It is of some consequence to have facilities to reach navigation for shipping the product and receiving a supply of fuel. But the one prime consideration on the part of experienced men is, to plant the derrick on a spot directly on a line between two paying wells, where there is doubtless the best chance of striking a good vein. Indeed, without breaking ground, the managers, whose well-springs are thus threatened, may take the alarm, and offer to buy out the new-comer at his own price. For this is one of the methods by which operators sometimes make their fortune.

A spot having been selected, the next business is to get an engine, erect the engine-house, the derrick, and other outworks. The house is a simple structure of rough boards, with perhaps a bunk for the engineer's sleeping apartment. The derrick stands at the distance of thirty or forty feet, the walking-beam, which plays upon a heavy upright pillar, called "the samson-post," stretching between them. The walking-beam is a heavy timber, from fifteen to twenty feet long. The derrick is a sort of pyramidal structure, resting on a square base, each of its sides being from ten to fifteen feet long, and rising to a height of forty or fifty feet, the summit approaching a point. The four principal pillars are strongly laced together by cross-timbers, into one of which rungs are driven, to make it serve as a ladder. Occasionally the whole structure is covered with boards, making it look like a tower; but more frequently they are content with protecting the driller from the elements. Under the apex rests a pulley-block, through which passes the long and powerful cable used in the work. The object in making the derrick so tall is to enable the workmen to use a longer and heavier drilling apparatus than formerly; its various parts, when put together, forming a continuous iron bar of thirty feet in length. It also enables them to withdraw or put down the tubes more readily. The weight of the tools now used in drilling commonly exceeds one thousand pounds, striking a powerful blow at each revolution of the crank.

The next step is to put down the surface-pipe or driving-pipe to a sufficient depth, so as to prevent earth or stones from falling into the pit, either while drilling goes on or afterward. This is a work both of difficulty and delicacy, since the pipe must be forced down through all obstructions to a great depth; while it must be perfectly vertical. Sometimes a hard boulder is encountered below the surface, which bends the tube to one side, in which event the work has to be abandoned. The depth to which operators usually force this down varies, many striving to drive it some distance into the first sand-rock, while others content themselves with reaching twenty, thirty, or forty feet. The pipe, made of cast-iron, has commonly a five-inch aperture and is one inch thick, being cast in lengths of nine feet. The apparatus for forcing it into the ground is a pile-driving machine, very simple in its construction and mode of operation. A wooden wheel and axle, termed the "bull-wheel," is geared to the engine by means of a stout rope, which, in the hands of one of the workmen, can be made tight to the axle or let slip at any moment, the rope being wound four or five times around it. When the engine and crank move, up rises the ram, a block of four hundred or five hundred pounds, to a height of perhaps six feet, between strong wooden shears to keep it in its place; the rope is then let slip, and the ram descends with a stunning blow on a plate or cap placed on the top of the pipe. At every blow a perceptible movement of the tube downward takes place; and it has sometimes happened that the driving-pipe could be put down its required length in little more than a day; in very hard ground, however, the operation may consume a whole week. The object of the bull-wheel, to the outside of which hold-fasts are nailed at short distances apart, is to operate as a brake, in the first place, so that a downward motion can be checked at any point in operating the well, and also to enable the workmen to raise or lower the tools, tubing, etc., in case the engine should not be running. By the aid of the pulley above, and the leverage of the wheel below, one man will, using hands and feet, bring half a ton up from the nether regions.

Sometimes, indeed, the pile-driver proves unable to force the driving-pipe to its desired depth, owing to a bed of sand-rocks intervening. In this case, it is customary to set the drill at work, making a three-inch opening through the obstruction. This creates a vacuum, into which the fragments broken off by the pipe in its descent can fall, when the ordinary appliances above are apt to force it downward. Otherwise the men will be compelled to pull up stakes and try elsewhere. Colonel Drake was the man who introduced this implement into the oil regions, as also sundry other valuable improvements.

Now begins the task of drilling—a task requiring from four weeks to as many months. A strong cable is coiled round the bull-wheel axle, and passed through the pulley-block at the top of the derrick, its other end having attached to it, by strong clamps, an instrument called "the temper-screw." This is ordinarily about three feet long, a thread being cut into it for half its length. As the chisel used in boring is indirectly connected with the screw, it will be seen that by turning the latter on its stem the former can be elevated or depressed at pleasure, until the end of the thread has been reached, when it becomes necessary to open the clamps and attach the instrument to a different point on the cable. This turning of the screw has to be done by hand, instead of being regulated by the engine, on account of differences in the hardness of rocks, some allowing a full revolution at every few blows, others only once in five minutes.

To the lower end of the cable are attached successively "the sinker," "the jars," "the auger-stem," and "the centre-bit," "drill," or "chisel," the whole being of iron or steel, and nearly thirty feet in length. The auger-stem is about fifteen feet long; the sinker, eight feet; the jars, three feet, and the drill, two and a half feet. The jars consist of two slender iron bars, so framed as to slide into each other fifteen or eighteen inches. The object of this instrument is sufficiently indicated by its name. It happens frequently that the chisel gets fast in a crevice or mud-vein, while the cable, several hundred feet in length, stretches so far that it is impossible to communicate to the drill that jerking motion which might extricate it. But the sides of the jars having, by the weight of the sinker above, been pressed into each other, like those of a telescope, the raising of the cable imparts a jerk to the tools below, ordinarily sufficient to remove them. Not always, however; for I think it within the truth to estimate one well in every ten sunk as being idle this moment from the tools having got fast below. In some localities, one may hear of every sixth well having been abandoned on this account.

The drill or chisel is about thirty inches long, and two and a half inches in diameter, with a three-inch face. It is attached by a screw to the auger-stem, which is about equally thick, its principal object being to give weight to the blow; while that of the sinker, as already stated, is to cause the jars to slide into each other. The cable is a stout inch-and-a-half rope, and must be sufficiently long to pass over the pulley and reach down to the bottom.

The engine—usually a portable one, giving from eight to fifteen horse-power—having been fired up, makes its first revolution, communicating, by a crank, motion to the walking-beam, which, in turn, moves the cable and the drilling apparatus. The driller takes his seat above the devoted spot, adjusts the chisel to it, and down it descends, striking thirty or forty blows to the minute. Between the strokes it requires to be kept moving round, to make the opening uniform and prevent accidents to the tools. With this there is also combined a slight downward motion, every few strokes, by a turn of the temper-screw. One blow tells the whole story. Hour after hour, day and night, the drill keeps churning up and down, punching the rock, and accomplishing from one to six inches per hour, according to hardness. At intervals, the apparatus is raised, and the centre-bit taken off, to be replaced by the "reamer." This is a round bar of iron, faced with steel, the centre and sides of its face being more or less hollowed, while its length (usually four and a half to five inches) corresponds with the proposed diameter of the well. The object of this instrument is to smooth the sides and widen the orifice to its proper size. In working, it requires to be kept turning and sinking as does the chisel.

The reamer is followed by the "sand-pump," an iron or copper tube about five feet long, with a valve opening upward in its nether extremity. On lowering this pump into the pit, the valve opens, and the pulverized rock, reduced to a pulpy mass, is drawn into the tube, hauled up to the surface, and discharged.

The sand-pump occasionally brings up, from the first sand-rock below the surface, small quantities of oil; but his is apt to receive no attention, operators knowing that it will soon exhaust itself. More frequently a good vein may be struck in the second rock; but on the lower part of Oil Creek and on its tributaries, this is usually allowed to pass unheeded. On the Alleghany and on French and Sugar Creeks, they seldom drill further than through this stratum. Elsewhere, on entering and passing through the third layer, "a good show" is most eagerly looked for, as the sand-pump comes up, filled with the gray, sloppy mass from beneath. If this begins to turn darker in color, separating into a thick, heavy sediment, which settles on the ground, and a green, slimy liquid, which floats away toward the nearest hollow, great indeed is the rejoicing; for the prospect of one hundred thousand dollars is within view. Indeed, the vein struck may, in an instant, anticipate reamer, sand-pump, tubing, and every thing else, sending up a spirt of petroleum which shall smite the top of the derrick and drive away the workmen, its rage only cooling sufficiently to permit them, after the lapse of a day or two, to return, insert the tubes, and guide the generous overflow into the cistern.

Examples of such inordinate zeal and energy are, however, very infrequent—certainly not more than one in a hundred, especially in localities which have been already well bored and pumped. The last instance that occurs to my memory was near Petroleum Centre, the freakish well being situated high up a ravine, far from its fellows. It is supposed to have discharged two thousand barrels before being brought properly under control. Much oftener the "show" is followed by an uprising of salt water, with perhaps only a few globules of oil to the barrelful. This water must be exhausted before a better quality of liquor makes its appearance; and in some cases days and even weeks have been spent in the process.

Setting aside such extreme cases, however, we will suppose that oil is actually got in paying quantities. The next process is to tube the well; though many prefer first to sink it some feet deeper, in order to furnish a receptacle for sand or gravel that may be washed in by the newly opened veins. To put in the tubing successfully, it is not only necessary for the superintendent to have some general knowledge of the mode of operating, but the characteristics of that particular locality. A mistake in this respect is apt to be attended with serious losses of time and money, some wells having had to be re-tubed as often as forty or fifty times within six months. The most experienced managers express the belief that large numbers of wells are totally unproductive from defective tubing. The tubes, of iron, are usually two inches in diameter, by about fifteen feet long, the ends screwing into each other. These are raised somewhat, in order to give greater strength and furnish a hold for "the tongs." To the lower end of the first joint is attached "the chamber," a copper vessel containing the pump-valves. This being got ready, the tube is hoisted upward by the tackle, and let down into the orifice. An instrument called "the pipe-tongs" lays hold of the tube at the joint, and prevents it from falling in till the second section is screwed in, when it is lowered further. This process continues to go on, section by section, till the work is completed. Operators differ as to the depth to which the chamber should be lowered; one of the most successful in the oil region says it ought to go within twenty-five or thirty feet of the bottom. Three or four men can put down five hundred feet of this in a common working-day. The weight of such a continuous pipe is between eighteen hundred and two thousand pounds, all depending from the tongs, which rests on cross-pieces of timber at the opening of the well.

By accident or design it has sometimes happened that the clasp of this instrument has been opened, letting the tube drop to the bottom, and occasioning much loss as well as trouble to recover it. For this purpose two instruments may be made use of. One is a short iron bar, made sufficiently sharp at the lower extremity to enter the tube and slide some distance into it, attaching itself to the inside surface by means of "steel dogs." The other is somewhat trumpet-mouthed, so as to slide over the end of the tube, and catch hold of its outer surface in the same manner. But in neither case is success certain to follow; and the manager may have the mortification of finding all his efforts followed by the abandoning of the work.

Before the well usually takes to flowing, and always before it can be pumped to advantage, it must be "seed-bagged." The object of this process is two-fold—first, to prevent the escape of gas upward, except through the tube, where it will assist in forcing up the oil; and second, to prevent the descent of water or gravel from above, which would have to be pumped up again at almost infinite trouble and expense. It will readily be seen what a heavy pressure is thus brought upon the bag which supports a column of water from two hundred to four hundred feet high, as also upon the tube itself. To effect this, a strong leather bag, shaped somewhat like a boot-leg, is filled with flax-seed and drawn over the pipe, along which it is made to pass down to the required depth, having been first made fast to a joint—the lower end of the bag very securely; the upper rather loosely, so that, in the event of having to withdraw the tubes, this fastening may break, and the bag turn over and inside out, spilling the contents and letting down the column of water. When the flax-seed gets soaked it bursts, swelling to such a degree as completely to fill up the vacant space between the tube and the outside wall, putting an effectual stop to all movements in that quarter. But it sometimes happens that the seed-bag bursts, in which event three days are apt to be lost in replacing it and pumping out the surface water which accumulated below. Experts usually place the seed-bag between the second and the third sand-rocks, if oil is got from the latter, say at the depth of three hundred and fifty feet; but in the event of its coming from the second, then at such a distance from the surface as will shut off all the fresh water without interrupting the flow of an oil-vein. Mr. Bliss, an experienced and successful operator, puts in the tubing first, without the seed-bag, and pumps out all the sediment at the bottom. He then puts in the bag; and although this involves a little loss of time at the outset, he observes that he has never had to seed-bag a second time, as others often do. Thus equipped for duty, if the well does not evince a disposition to flow profusely, the pumping machinery is set at work. The pump-rods are of wood, their ends fastened together. This simple apparatus is connected with one end of the walking-beam, already described; an iron pipe is also joined to the upright tube, to receive the petroleum and convey it to the tank at any required distance or situation. The first movements require to be made cautiously, until it has been ascertained that every part is in working order, when more steam is let on. Up comes the brine or the oil, but more commonly both, the former in much larger proportion at the outset.

The tank or cistern is a circular vat, made of wood, and having a capacity ranging from two hundred to twelve hundred barrels. The custom formerly was to make this vessel square, securing the timbers by heavy uprights; but its greater tendency to leak has made that give place to the circular form. When wells discharge copiously, it is customary to erect whole streets or blocks of these enormous vessels, covering them with a long range of roofing. For purposes of shipment, it is desirable to get them convenient to the road or some navigable stream, which is sometimes lined with the black, slimy monsters for long distances. The tanks are connected together by a system of pipes, which distribute the liquid from one to another. Near the bottom of each is inserted a stop-cock for letting the salt-water escape, after settling there by its greater specific gravity. In spite of all precautions, the leakage of petroleum through the wood, or its waste in filling into barrels, is considerable, so that on many farms it would pay well for collecting. Some managers have caused pipes to be laid down, by which vessels can be laden in bulk—a mode of preventing loss in more respects than one. In the flowing wells, little or no water comes up with the petroleum; in the pumping wells, the two may be found in all imaginable proportions. The brine also differs in strength, being in some cases equal to that found at Syracuse, New-York, and in others tasting slightly brackish. It differs also in some localities, according to the season, being weakest in spring, when the fresh surface-water has found its way to the springs, by careless or malicious individuals and companies letting their works get out of order.

The cost of sinking and tubing a well of five hundred feet, which is the ordinary depth of those lately put down on Oil Creek, but is less than the average of wells on Cherry Run or Pithole, is between seven thousand and seven thousand five hundred dollars, unless the proprietors own the steam-engine or other part of the apparatus required. In this case, it will amount to between four thousand and five thousand dollars, a good-sized new engine costing twenty-five hundred dollars on the ground. At the beginning, wells could be bored for fifteen hundred dollars, and even one thousand dollars, including the outlays for engine and machinery, on account of the cheapness of materials, labor, and fuel, as also the smaller depth to which it was customary to sink the wells. If the work be now let out to a contractor, it will cost from five to seven dollars per foot, in addition to outlays for derrick, engine, and other fixtures. If workmen be engaged by the day, first-class machinists, carpenters, blacksmiths, etc., must be paid five dollars per day; drillers, about four dollars; engineers, about three and a half; and common laborers, from two to three. Two sets of drillers and engineers usually relieve each other, the machinery running the full twenty-four hours. Teamsters command from thirty-five to fifty dollars per month, with board and lodging; the other figures suppose the employés to provide these for themselves. The cost of horse-hire is very heavy—in winter, enormous—the charge for a double team and driver being at least ten dollars per day. The bituminous coal made use of, though it might be obtained from hills not ten miles off, comes from Pittsburgh and costs from sixty cents to one dollar and twenty cents per bushel, according to the season and the distance of transportation by wagon.

To obviate the last-named outlay, now becoming oppressive, Mr. Wade, superintendent of the Lady's well, applied the escaping gas to the generation of heat with the most triumphant success. Large numbers of engines are now driven by this carbonetted hydrogen, which is, without difficulty, conveyed from the tube into a large barrel or cistern, and thence into the furnace. Nothing could be more beautiful, economical, and perfectly safe than this arrangement. The few splinters of wood thrown upon the grate in the morning to light the fire, serve also to keep it agoing all day. The gas is used to give light in the engine-house at night, and has, in at least *one* instance, been conveyed to a private dwelling, where it is used for ordinary culinary and heating purposes. Such economy has hitherto been a rather novel feature in the management of affairs in Petrolia.

On the other hand, a similar experiment made at the Auburn well, on Cherry Run, belonging to the Cherry Valley Oil Company, resulted unsuccessfully, diminishing the daily flow of sixty barrels to less than one half of that quantity. The attempt was consequently abandoned, when the former flow returned. The company have since then inserted a clause in all their leases, prohibiting those who put down wells on their premises from using the gas as fuel.

To the best of my knowledge, this is the only instance in the oil region of Pennsylvania, where such an effect has followed the introduction of gas as fuel, whether taken from pumping or flowing wells. On making inquiry on this subject, from the manager of one of the latter class, I learned that they had made repeated experiments, and in no instance with a diminished yield of oil. But in this case there was a *very large volume* of gas escaping; while the production of oil was far below that of the Auburn well. In the latter, the gas appears to have been barely sufficient to expel the liquid, having so little surplus energy to spare that its capture and imprisonment proved too much for exhausted nature. The propriety, therefore, of using gas as fuel must, it would seem, depend upon its quality and strength relatively to the discharge of oil, and perhaps to the time during which a well has been in operation. In some instances, a single tube supplies gas to two, three, four, even six engines, without any apparent detriment to the yield of oil.

It has been stated that along two small streams the wells have to be sunk deeper than on Oil Creek. The difference varies from fifty to two hundred feet, and is due solely to the higher elevation of those bottoms at the place of carrying on operations. To reach the same rock on any of the tributaries of Oil Creek, it will, indeed, be necessary to make allowance for this elevation, adding or subtracting more or less for the natural inclination of the rocks. Some operators, it is true, report that in the wells planted against the bluffs or on the table-lands, it is *not* requisite to sink as many feet deeper as the difference in surface elevation would betoken. I reply that this statement is partly made with a view to selling lands or wells, and partly to mistaking the second rock yield for that of the third, the former being unusually copious for these times on account of the comparatively great distance from other wells. It is possible that on the uplands the bed which corresponds with the *first* sandstone of the valley will in places yield copiously for a time.

On passing down the Alleghany River, and entering the valleys of French and Sugar Creeks, owing to the southern dip of the rocks a depth of between one hundred and two hundred feet additional must be sunk before striking the third layer. So far the custom there has been to stop at the lower margin of the second stratum, or between four hundred and five hundred feet down. The petroleum coming from that bed is everywhere better in quality than that obtained lower down; and on French and Sugar Creeks is of the rich, heavy, and high-priced kind known as lubricating oil. In its crude state, it sells for many times the price of the illuminating variety; but is never got in such large quantities as the latter. The Alleghany oil readily commands one dollar per barrel more than that obtained from the third sand-rock on Oil Creek or its tributaries.

During the past twelve months, derricks have been erected and boring has commenced in the various rivulets discharging into the principal streams; but in some instances also operations have begun on the open table-land, at elevations of two hundred or three hundred feet. That the uplands will yield abundantly I have no manner of doubt, the veins there being probably quite as numerous and productive as those in the river-bottoms. Whether the existing appliances will suffice for wells from three hundred to four hundred feet deeper than the present remains to be seen; but if not, American ingenuity will doubtless remove all obstacles in the way of the further development of the upper region.

Allusion has been made to the loss of tools by striking crevices in the rocks. Some experienced men regard this as a happy omen of the early discovery of petroleum; since it is usually found in such caverns or pockets, which are believed to have originated in geological dislocations. It is supposed that they are connected by subterranean channels, so that when one is penetrated a whole system of rich springs will be opened. Experience does not fully bear out such expectations; and hence the loss of tools is apt to occasion bitter regrets rather than congratulations. Above Petroleum Centre, a mud-vein is apt to be struck in the middle of the third sand-rock; and so frequently have the tools got fast in it that it has become customary to stop short there, instead of boring through it. The depth of wells there will, accordingly, be found on an average fifty feet less than lower down the creek, where this obstacle does not present itself.

A rough estimate of the whole cost of sinking a well has been given above. Making some allowances for "the chapter of accidents," it will be seen that the average outlay amounts to about fifteen dollars per foot, distributed nearly as follows:

Engine and boiler of fifteen horse-power, delivered,	$2,500
Engine-house, derrick, bull-wheel and cable, samson-post, walking-beam, and fixtures,	750
One tank, five hundred barrels,	250
Drilling-tools, ram, driving-pipe, tubes, etc.,	1,000
Labor, fuel, and extras,	2,500
Breaks, delays, and other accidents, say,	500
Total,	$7,500

These figures may perhaps be reduced one half, if the operators own the engine, and put forth their own hands freely in the various operations.

CHAPTER IV.

"STRUCK OIL"—THE LAW OF LAWLESSNESS.

AT last, the labor, the anxiety, the profuse expenditures of long months have been crowned with triumph. The sand-pump has brought up unmistakable evidences of the existence of petroleum, which floats upon the surface of its load of mud in long, dark streaks, most gratifying to behold. The well in question is one success, after two, or it may be *ten*, failures, and gives promise of paying all expenses connected with the whole number. A grand halo of wealth and beauty thenceforth rests upon every adjoining field and patch of wood, which has become in some measure *sacred*. That which was almost valueless yesterday is viewed as a princely inheritance to-day. The omnipresent telegraphic wire has already caught up the joyful intelligence, and whispered it to all the outskirts of Petrolia, as also to interested parties in Wall, Chestnut, and State streets. Before the setting of that day's sun, an excitement will have sprung up in all parts of the oil region, and crowds of speculators, operators, and curiosity-seekers will have been on the way to a spot which exceeds all the dreams of poets or relations of prophets about Parnassean and Castalian springs, fountains of perpetual youth, philosophers' stones, miraculous blessings of widows' cruses, etc. One well, with its one thousand or even its one hundred barrels per day, would justly have been regarded as a world's wonder; *this*, being a matter-of-course affair in Petrolia, receives the attentions of few except those who propose to make money out of it. What is "common" not only ceases to be wonderful, but becomes in a measure "unclean." If, however, in the realms of poetry and miracle there be little or no interest, in that of solid fact there is abundance to make up for the deficiency. As the greasy liquid finds its way into the tank, the achievement occasions a triumph to the workmen, brings fortune to its owners, attracts fresh millions to the country, and furnishes pabulum for excitement to all. The modern miracle is thus a sand-pump, which may bring up a million of dollars at a single ascent.

But here, at the outset, comes in "the law of lawlessness." For precisely as every human being has his own set of features, tone of voice, and the like, so each individual oil-spring has *its* characteristics, with regard to the escape of water, gas, petroleum, or all three. In one, the flow of the last-named will be continuous and uniform, day and night, not varying more from week to week than a spring-brook. It may or may not be accompanied by a large escape of gas, visible to the naked eye, though it ordinarily is by some. In such a flowing concern, there is no noise, except the splash of the liquid falling into the tank. But in others, a regular periodicity takes place in the discharge. The Coquette well, for example, emits a succession of sounds as loud and sharp as the exhausts of a small steam-engine, and occurring in tolerable order every ten seconds, in such a manner as 1, 2, 3, 4; 1, 2; 1, 2, 3; 1, 2, 3, 4; sometimes two of these coming off together by an extraordinary effort. A copious discharge of gas comes off with each of these eructations, ascending from the tube like a thin smoke, to mingle with the atmosphere. The Wild-Cat well, at Petroleum Centre, and the Yankee, on Cherry Run, remain silent for forty and twenty minutes respectively; then begin to foam and spirt, the oil coming off at first only in drops, but increasing by degrees until the tube belches forth quite freely; these discharges afterward decrease in violence, and finally stop altogether, after the lapse of from five to eight minutes. Each escape of liquid is accompanied by a sharp report, which may be heard one hundred yards off or more. Both wells have been some months in operation, and belong to the second class, their yield being under one hundred barrals per day. Those which have a continuous flow are apt to produce more abundantly, as do the Reed and the Mountain wells, on Cherry Run, and the Craft well, on Bull Run. Yet this rule is not universal; for the Coquette, discharging spasmodically, is the most productive in that whole region; while a well on Cornplanter Run rather drips than flows one or two barrels per week.

Others again are found to remain quiescent for twenty-one or twenty-two hours in the day; then to break forth in one continuous flow, or a succession of belchings, for the other two or three hours. A few run for six hours, and then subside, or distribute their favors over twelve hours in the twenty-four. The Dunn well, on Watson flats, produces freely from morning till midday; then the supply diminishes or stops altogether, for the rest of the day, the pump bringing little but salt water. But in the case of the pumping-wells, the rule is pretty common to spirt forth from a few drops to several quarts at each revolution of the engine, varying according to the season, but not modified by day and night.

No person has yet arisen who can explain the full *rationale* of these vagaries on the part of Nature. No one cause can be assigned which will account for all the phenomena. There can be no doubt that the efforts of the imprisoned gas to escape into the upper atmosphere have much to do with forcing up the oil; yet in numerous works of the largest flow the escape of gas is comparatively small, and altogether insufficient to carry such a column of liquid up the tube. For it is needless to observe that a discharge of three thousand barrels per day would almost completely fill a two-inch tube, making in it a nearly unbroken column for the height of five hundred feet. Would the comparatively small body of gas escaping therewith be sufficient to force upward such a load, exerting a pressure that is practically unknown in hydraulics?

In the cases of spasmodic wells, the gas may have a great deal to do with bringing petroleum upward—in many, there can be no earthly doubt on the subject. While visiting the Dunn well, the superintendent conducted an experiment that proved this point satisfactorily. They had been using a portion of its gas in the furnace to generate heat, but on turning it *all* on the fire, the discharge of oil and water stopped instantly. After a few minutes, the gas was partly permitted to escape as before, when the yield began to return to its former amount. In plain English, the energy of the gas coming up the tube was insufficient to do its work, and bear the additional pressure upon it necessary to its confinement with a view to use. While on this subject, I may mention another curious experiment made by the same manager, (Mr. Morrell,) who conducted this gas through paper surrounded by *ice.* The result was that the gas was converted into a white, wax-like substance, which settled on the paper, and is supposed to have been paraffine.

Nor will the theory that the oil and briny veins are supplied from springs higher up in the hills, suffice to account for the phenomenon. For, in the first place, there is no evidence that such elevated sources exist close by in quantities sufficient to feed such wells as the Philips, (flowing nearly four thousand barrels daily for a considerable period,) *and force it up to the surface.* The origin of most of our water-springs is easily accounted for by percolation through the under-strata from higher levels; but in the case of Artesian wells, constructed on such interminable plains as those of Texas, this theory proves insufficient to account for the outburst which has followed. Whatever the agency or mode of operating on Nature's part there, in my judgment, the like has been at work in forcing up much of the rock-oil.

Even if, as in nine cases out of ten with productive wells at first, and in all ultimately, an application of the pump be required, the same lawlessness is found to prevail as does in the flowing wells. We can calculate with certainty upon nothing in the oil-basin; we need feel surprise at nothing—not even if the yield should be Orange county milk. The first revolution of the engine may bring up petroleum or salt water of any imaginable degree of strength. Nay, in a few instances, *fresh* water has come up with the former. One of these was near the famous Sherman well, where the two liquids had to be separated by artificial means. The case of the Sherwood and Kelly well, on Cherry Run, which threw up three thousand barrels a day of fresh water to the height of sixty feet, flinging out fragments of rock several pounds in weight, is not so remarkable, as this vein was struck at the depth of seventy feet, and may have been fed from the hills above. There was no oil with that water, and I refer to the circumstance to show one of the obstacles that must occasionally be encountered by operators.

It may require days or weeks to exhaust the heavy liquid (brine) before the lighter comes within reach of the pump. In one instance, the proprietor having got the impression that oil certainly existed, kept on pumping for *six weeks* before his efforts were rewarded with success. The best well (Ingersoll) now in operation on the Watson flats was incorrigible until nearly the whole winter months had been exhausted in forcing up the salt water; at last, it also yielded to perseverance, emitting a stream of from fifty to seventy barrels daily. There is every reason to believe that hundreds of works have been abandoned by those in charge of them, because, after pumping a few hours, the petroleum did not make its appearance; yet the perseverance which wrenches victory out of defeat might have made them productive. For it is one thing to be flooded with salt water at the outset, before oil has been reached, and another to keep on pumping a well which has gradually declined from a high figure to nothing, in spite of re-tubing and all the other known appliances.

And why, except from a disposition on the part of Dame Nature to indulge in freaks, should water be found floating on the surface of oil, the latter being the less ponderable of the two; while gas, coming up with it, is *im*ponderable? Various explanations have been offered; yet none, in my opinion, that will account for the regularity of this phenomenon. It has been supposed that, in consequence of carrying the lower end of the tubing below the level of the petroleum, the latter cannot enter the orifice until the former, lying beneath it, has been expelled. This crude theory is confuted by universal experience; for it surely could not help happening sometimes that the tube would be so conveniently short as not to pass beyond the petroleum. Besides, if substances arranged themselves in the well according to their specific gravity, would not the gas make its appearance first, instead of its awaiting the removal of the water, and then assisting to expel the more valuable liquid? Once more, the sand-pump shows that, in numerous instances at least, the salt veins are reached in boring rather earlier than the oil veins; though I concede that the testimony on this point is any thing but explicit or uniform. But it occurs often enough to refute the notion that the oil actually reaches the top in the wells and settles there.

Another explanation has been offered, namely, that a vein or veins may be struck at a point below the general level, where they are fed from reservoirs somewhat higher up; hence that, inasmuch as the brine would settle to the bottom in the natural crevices, it must be pumped out prior to the oil and gas being able to come downward and forward. Further, that it is by tapping the source at the *right* place that such immensely productive works as the Philips, the Noble, the Sherman, and the Coquette have begun to produce at once such prodigious quantities of oil, and that almost entirely free from water. This theory is much more satisfactory than the former; yet it will not suffice to explain the regularity—almost uniformity—of brine being found first in order. If, in the multitudes of experiments made, petroleum came first as often as does the salt water, it would not be necessary to look further. Unfortunately for it, such is not the case. It does not appear reasonable that the lower reservoir should be almost invariably struck by borers, and the upper scarcely ever. Some other solution of the difficulty must be sought to cover this unexpected regularity in the heaviest substance coming up first—one which will apply to the ninety-nine cases as well as the hundredth. Elsewhere I have alluded to Mr. Morrell's experiment of separating oil and water in the atmospheric pump; though whether this will apply when a quantity of gas is mixed with the liquids, I am not advised.

And whence the salt that, in the form of brine, gushes upward from depths of one hundred or one thousand feet? Those who entertain the theory that petroleum has been distilled from the upland coal-beds which once existed along the present oil-basin—the products percolating through the different strata to their present depths—account for the existence of salt-springs by the fresh water filtering downward mixing with muriate of soda in sufficient quantity to form the brine. But to account satisfactorily for depositing in the rocks the quantity of muriatic acid necessary for this purpose, is every whit as difficult as to answer the original question. It has been only removed one stage further off. The salt water has its position in the scale, as has the fresh water its assigned position or limit. Above the first sand-rock, the water is invariably fresh; between the first and the second, it is expected to be such; below that stratum, it is as certainly expected to be brackish or briny, the degree probably depending on the proportion of fresh water let down the orifice, while the work of drilling has gone on. As a rule, the surface water does not go down through the second rock, or the brine force itself above it, until an artificial opening exists, the exceptions being so few in either case as to establish the general principle here laid down. Besides, we all know that water, in passing through sand or clay, is apt rather to part with impurities than take others up.

The only satisfactory explanation of the existence of so many salt-springs is, that the strata in which they abound, at one period in the world's history, formed parts of the ocean-bed. This may have consisted of limestones, sandstones, or conglomerates, all saturated with brine, and reposing on what had been beds of clay which contained carbonaceous ingredients; while the process of baking this clay into shales filled it with cracks and seams, that have since become so many veins, filled with salt water or petroleum expressed from the rocks subsequently formed; the whole being upheaved to their present elevations. But who shall fill up the picture of those ages, of which this is scarcely an outline?

The relationship between brine and petroleum is intimate, yet they are not invariably found close together. Where oil is got in the second sand-rock, it has sometimes happened that the "show" of salt water was scarcely perceptible in or above the oil vein. By sinking a few feet deeper, however, the brine would flow up so profusely as to completely monopolize the tube and choke off the more desirable product. An instance of this happened in Franklin, involving the immediate ruin of a profitable well. Another disaster of the same kind, close by, was only prevented by forcing down a plug, which stopped any further uprising of the salt water. But as a rule, the two are found in such close contiguity that it is difficult to decide which has been sooner reached. At Tideoute, the alliance is so intimate that it has come to be considered a maxim: "No salt, no oil." As to the why or wherefore of this connection, our superficial philosophy must place its finger on the lip, and be modestly silent—for the present.

Nor does the law of lawlessness end here. As men have been accustomed to measure the decreasing temperature, in ascending from the sea-level, so have they measured its increase downward, as found in mines, caverns, and the like. This increase is about one degree Fahrenheit for every sixty feet. At this rate, we should find the oil, water, gas, etc., brought up from a depth of six hundred feet, to be (say) ten degrees hotter than they are ordinarily found on the surface. My attention was first called to the contrary by the superintendent of the Cherry Valley Oil Company, who observed that from the Yankee well oil flowed very little above the freezing point. In some few instances, whether owing to careless observations or Nature's freakishness, his remarks were not borne out by other operators, but in many more they were fully confirmed. Mr. Ostrom, of Titusville, who has been some years at work in the valley, took the pains to test this matter by the thermometer. He observes that different wells range through various degrees of temperature, according to the season. *In winter this is not apt to be so low as in summer*, when it is usually found as far down as forty or forty-five degrees; at the opposite season it rises, in some wells, up to sixty degrees, causing a smoke to ascend from the surface of the tank. The tests in all cases were made in a pail-full of liquid, as discharged from the pipe, in which the heat and cold would both be somewhat modified. The whole subject requires much more extended observations than have yet been given to it, previous to making which it is not advisable to theorize respecting the causes at work in producing this effect. The circumstance shows, however, that men who have undertaken to calculate at what distance from the surface our globe is in a liquid state, have reckoned altogether too fast.

Indeed, as if with a design to nullify theories or teachings of any kind as to the subterranean world, and absolutely to bewilder the investigator, rendering him hopeless of arriving at any general conclusions, we have the fact vouched for by multitudes of experienced operators, that *wells yield much less freely in winter than in the summer season.* This assertion is also called in question by some practical men, who seem to be as diverse in their relating matters of fact, as Nature is erratic in carrying on her processes. Thus, an intelligent superintendent at Petroleum Centre admits the fact of a difference to a slight extent, but denies that it should be attributed to the cause assigned, namely, the agency of or sympathy with conditions existing in our upper atmosphere. His belief is, that the collection of paraffine in the tubes, which is much greater during cold than hot weather, and prevents the free passage of oil, water, and gas, creates all the difficulty. As proof of this theory, he alleges that the paraffine collects chiefly on that part of the tube which is near the surface and thus exposed to the atmosphere. Elsewhere I was assured that they had "tested" this explanation by re-tubing in winter; but invariably they found the same decrease at that season. They pointed also to the Noble, the Fountain, and other flowing wells, which suspended action altogether in winter; while the pumping-wells on whole farms had been found so unprofitable then that their managers either stopped altogether, or kept on to prevent the supply of petroleum from stopping for all time. Some managers report a difference of thirty per cent between the average production of wells in summer and winter. The superintendent of the Briggs Oil Company declares he can predict changes in the weather for twenty-four hours in advance by observing the yield of his wells. On the approach of a cold rain or snow-storm, they begin to slacken up a full day beforehand. On the other hand, after commencing to yield more freely, it is safe to calculate upon pleasant weather. No better barometer, he alleges, could be devised to forecast the outbreak of storms than the decreased production of these long tubes, with petroleum instead of mercury. He thinks the are changes wrought by heat and cold, not the mere presence or absence of a certain quantity of moisture in the atmosphere. The Forest City well did not yield twenty-five barrels per day during the dead of winter in 1863–4, but recovered to an average of eighty-five barrels next summer. In December last it gave out altogether, and so continued till the eighth of March, when oil once more made its appearance, but in small quantities at first.

Testimony to the same effect is so abundant, that it cannot be upset by the experience, much less by the *opinion*, of a few individuals, no matter how trustworthy as observers and narrators. But the difference between summer and winter production is doubtless not so great in new as in old wells, or in flowing as in pumping ones. It is in that stage of decay, approaching exhaustion, where the gas coöperates less vigorously with the pump, that the greatest diversity as to season is found. But how or why such intimate sympathy should exist, when every means is used to exclude air from reaching the fountains, or gas from passing upward, except through the tubes, I do not pretend to be able to explain.

Perhaps one or two other observations may assist the reader in arriving at a better understanding of the subject. The localities where this aërial sympathy is most acutely felt, contain large numbers of old and idle wells. In those which have been sunk on ground least punctured by the drill, I believe there is the least cause for complaint. Still, at the Economy wells, near Tideoute, where the land has been little bored on that side of the river, there is an acknowledged difference of fifteen or twenty per cent in favor of the summer season. The veins supplying those wells have certainly not been reached by others within the distance of half a mile.

Mr. Fox, manager of the large interest owning Petroleum Centre, observes that sometimes wells flow pure water, then oil alone, then water only. In that locality, he believes the oil and water are got together below the third sand-rock.

Even with regard to the veins of petroleum, such a degree of uncertainty exists, in advance of actual trial, that they may be justly placed among the objects or phenomena subject to the law of lawlessness. Before sinking a well, it is utterly impossible to foresee whether it will or will not strike the oil-vein that feeds a neighboring well. Scores of instances might be brought forward to prove that a constant and intimate sympathy below ground *does* exist between wells situated hundreds of yards apart. On the celebrated Tarr farm, several weeks were consumed last spring in pumping out the water which had collected from above, in consequence of withdrawing the tubes and leaving the orifices open. This water passed from one well to another, flooding those on the whole farm. The same costly process has to be gone through with the present season, many, at the time of my visit, having just begun to discharge the bilge-water, which had been collecting during the winter months. The great Philips spring was reached by a rival interest, which carried off two thousand barrels a day, and nearly dried up the original source. By withdrawing the tubes, each manager found he could render the other work utterly unproductive. Finally, after much litigation and delay, the difficulty was settled by each proprietor agreeing to give the other one third of his own yield, on condition of the other keeping his tubes in proper order. In one case, this connection was so intimate, through subterranean channels, that the sounds made in drilling one, though not audible at its mouth, could be heard coming up the next well. On the Watson flats, a torpedo, let down nearly five hundred feet into an exhausted well, exploded and started the oil in another, though nearly half a mile distant, improving the character of the one subjected to the operation. On the other hand, Mr. Duncan, an experienced operator, assured me that one of his wells, which had been sunk close by the boundary-line of his lot, yielded so largely as to excite the cupidity of a neighbor. The latter decided to bore at a point within eleven feet of Mr. D.'s well. After sinking some distance, an offer to pay all expenses then incurred, on condition of abandoning the attempt, was promptly rejected. The work was carried down to a point below the Duncan well, but it yielded nothing. Not disheartened at this failure, the intruder made a second attempt at tapping the spring, going a few feet further off and in a somewhat different direction; the result was a yield of about one barrel per day for a time, which led him to abandon the enterprise. Many other instances of the same sort might be adduced, showing that the notion entertained by some of a grand reservoir of oil lying below ground at a given depth, is unsupported by facts. Even a well sunk in a direct line between two producing works does not invariably strike the common vein; for though passing within a few inches of the spot, a bed of clay or solid rock may forbid the exit of petroleum.

"Surface indications," it has been shown, relate simply to the oil-springs found in the upper sand-rock, and these are commonly so insignificant that operators pay little attention to them. To reach the more abundant sources in the inferior rocks there is no certain guide, the only approach to such being to get as nearly as possible on a direct line between other producing wells. The law of lawlessness also applies to *the quantity* of oil which a spring will yield after being reached, as truly as to its prospects in advance. Two wells may be within fifty feet of each other, as closely as they can be operated to advantage; yet, while one pours forth a hundred barrels per day, the other will refuse to give out more than ten. Even their modes of contributing these quotas will vary, one flowing by fits and starts, the other emitting a steady stream in response to the pump. The one may run two years, while the other becomes exhausted perhaps at the end of three months or thirty days. Commonly, the contrast is not so sharp as this, wells on the same vein having a good degree of sympathy with each other; yet the purchaser ought to make the possibility of an entire diversity in time as truly as in yield, enter into his calculations as to productiveness.

In no other respect does the apparent capriciousness of Nature manifest itself more clearly than in the duration of an oil-spring. *The fact that wells would give out at last* was unknown, until it had been proved by abundant and most painful evidences. It is still called in question by some, who are either ignorant of the country or interested in creating misconceptions in the public mind. After so large a proportion of the works have gone through the several stages of flowing, pumping, deepening, re-reaming, etc., and then refused to contribute enough to pay expenses, it is unpardonable for persons at the East to be caught napping, either by investing their means in defunct concerns or expecting productive ones to last for ever. The excuses which a stranger is apt to hear or observe in print as to the number of idle derricks, are endless, and ought to arouse suspicion at once. "Oh! she was damaged by the flood;" as if a very profitable enterprise would be even temporarily abandoned for the lack of a few thousand dollars, which could be repaid within as many weeks. "She was never put down deep enough;" as if the apologist had been in the basement story, and examined how much further it was to the great petroleum reservoir. "She was abandoned in 1861, when oil had sunk in price to twenty-five cents a barrel, and has never been worked since;" as if crude petroleum had not since advanced to half as many dollars per barrel at the well. "Her lease has expired, and she is now in the hands of the original owners of the land;" as if they or thousands of others would not have rushed to the rescue of a productive concern, which was going to ruin for want of a little means. Rarely, indeed, are these excuses more than founded on fact, the truth in almost all cases being, that the wells have gone dry or were unproductive at first. The chances are believed to be quite as many in favor of sinking a new well as of deepening an old one, even though it may not have reached a greater depth than three hundred feet in the first instance.

Of those which were bored as early as 1861, I think it safe to estimate that not more than one in a hundred have continued to yield steadily ever since, and are now affording enough to pay operating expenses. The Economy wells at Tideoute are most remarkable exceptions in this respect. The fact that that spring should have been reached within one hundred and fifty feet of the surface, makes the "record" of those works the more interesting. Certainly the whole region of Petrolia has no parallel cases to these; and at this moment I do not recall a solitary instance where a well, no matter to what depth sunk, has continued to flow or pump, without serious interruption, for four years, as they have done. Let no lucky adventurer who may strike oil lay the flattering unction to his soul, that the spring which foams and bleeds so profusely in his presence will continue to pour forth oil in paying quantities for four, three, or perhaps even two years. Before the end of eighteen months, it will not only call frequently for repairs and renewals, but probably have settled down to figures so modest as barely to cover working expenses. An oil-spring insurance company, did such exist, would not issue a policy, upon the first discharge of a well, that it would hold out longer than twelve months; at any rate, it would be unsafe to accept the risk of its continuing to yield more than ten per cent of the average of its first month's production. Indeed, experts have assured me that they have set down *nine* months as the ordinary period of gestation; so that what additional contributions may be made are such as the lucky possessor has no right to expect.

But as this is a matter more properly belonging to the financial question than the idiosyncrasies of Petrolia, I leave its further discussion for another chapter.

CHAPTER V.

OBSTACLES IN THE WAY — THE MEANS USED TO OVERCOME THEM.

IN some form or other, obstacles to production present themselves from the day a well takes its place among existences till its last breath is spent. As the human system is subject to diseases, which, with increasing years and diminishing strength, augment in power, ultimately ending in dissolution, so with an oil-well. To some of these difficulties I have already alluded; but to be understood clearly they must be systematized and examined more fully under a separate head.

One of the first obstructions encountered is *the collecting of paraffine*, a substance whose qualities and uses will be found explained under the head of the refining of petroleum. It collects occasionally on the pump-rods, preventing their free passage up and down in the tube; a good well on the Miller farm had been thus thrown idle for some days, when I visited it. More frequently, however, especially during winter, this substance collects on the inside of the tubes with the same effect. In such an event, it is probable the adhesive element might be removed by heat, causing it to melt, run down to the bottom, and come up mingled with the oil. To accomplish this, it is only necessary to direct a current of steam down the tube for some time. As, however, the real cause of failure on the well's part can only be guessed at, previous to examination, it is customary to withdraw the tubes, unscrew them, clean them out, and afterward replace them. In a deep well, these operations will consume from two days to a week, according to the quantity of water collecting below. The task requires three or four men to perform it, and is altogether one of the least pleasant in that country.

Quite as frequently this waxy intruder is found *adhering to the sides of the well*, alighting, precisely where it ought not to alight, on the oil-veins, and sealing them up as tightly as if the operation had been performed by hand, with the use of sealing-wax. Here again the trouble is that managers, being in the upper world, can have no certain knowledge of the difficulty. As far as known there, the stoppage may be owing to the collecting of paraffine or to exhaustion in the supply. The tubes must once more be pulled up and examined. If nothing be found the matter with them, the supposition next in order is that paraffine has collected on the sides of the well. To remove this a tube is put down a long distance, and through it is directed a column of steam, which dissolves the adhesive substance and restores the power of the oil-veins. This consumes quite as much time as the other operation.

If this proceeding fails to accomplish the end desired, recourse is sometimes had to the *torpedo*, a thin cast-iron tube, four or five feet in length, and filled with gunpowder, which is fired by means of a galvanic wire. The object of this is to open or reöpen fractures in the rocks, and liberate quantities of petroleum locked up in them.

There is a good deal of trouble, as well as some outlay, connected with this device; and the experiments made with it have been too few as yet to justify any general conclusion respecting its merits. The fact that such an explosion should have started the petroleum in a well far distant, suggests the inquiry whether expenditures of the kind should not be borne equally by all companies and individuals in a neighborhood, whose wells require "rejuvenation"? One work is as likely to gain benefit as another from the operation.

When steam-pipe and torpedo alike fail to restore a well to its pristine vigor, or at least to a paying condition, and the manager has the quality of perseverance, it is customary to resort to other processes. The next step is commonly to sink to a greater depth, particularly if the well were put down, in the first place, less than five hundred feet. Along the Alleghany and on various parts of Oil and French Creeks, this operation is now in progress or contemplation at some hundreds of works. In the great majority of cases, however, it is undertaken by new interests, into whose hands the property has fallen by lease or purchase. By looking over the chapter on the statistics of production, the reader will find many instances where such endeavors have proved successful to a moderate degree—very seldom, however, in tapping a big spring lower down, after a tolerably good one had been exhausted in a higher stratum. When it is considered that in that chapter little mention has been made of unproductive wells, except to group them in mass, I am inclined to think that scarcely one well in seven—perhaps not even one in ten—ever succeeds in getting petroleum sufficient to pay the additional outlay incurred in deepening it. Certainly not one in fifty becomes a second-class concern, yielding from twenty-five to one hundred barrels per day, for the space of twelve months.

The other proceeding is to ream out the well, increasing its diameter from four to five or six inches. This has not been resorted to in nearly as many instances as deepening; but I think the experiments already made and the impressions of the workmen thus engaged, will bear me out in saying that it is not likely to be more successful than the other. The theory in re-reaming is that, being much more easily performed than boring anew, it not only removes accretions from the side of the well, but may lay open new springs or veins, separated from the boring by very slender partitions. Full and accurate statistical information, drawn from whole sections of the oil regions, is highly desirable. The trouble now is, that in this, as in nearly all other matters relating to Petrolia, the successes are blazed abroad by telegraph, newspaper, and private epistle; while the failures are glozed over, or at best only touched upon, as if they were matters of which the public must be kept in profound ignorance.

Among the many hindrances experienced is *the inefficiency of engines*, either by their getting out of order or having to stop from a deficiency of coal. Either of these drawbacks puts a complete stop to the production of petroleum, without making any diminution on the score of expenses. The engines first imported were almost literally "one-horse" machines. Their boilers were insufficient to generate the quantity of steam since found requisite to drive the pumps. The cylindrical part being usually from twenty-four to twenty-eight inches in diameter, and from four and a half to five feet in length, the power produced ranged between five and eight horses. Not that in every case a very large mechanical force is required for that purpose, the law of lawlessness applying to this as to all other departments of Petrolia. In some wells a pressure of ten pounds to the inch has been found sufficient for all purposes, in others six or seven times that pressure is needed. As the wants of a well cannot be foreseen, skilful operators now judge it advisable to enlarge the capacity and augment the strength of boiler and engine, which are, accordingly, now made sufficient to give twelve, fifteen, or twenty horse-power. On the properties of several companies one engine is geared to two pumps, or to one pump and a drilling apparatus. Near Titusville, a forty-horse engine has been introduced and set to pumping three wells, to which a fourth will shortly be added. The results appear to be perfectly satisfactory, as is manifest from the number of small engines which are for sale in all parts of the country. Indeed, considering the contiguity of numbers of wells on many farms with the high prices of labor and fuel, the marvel is that engines with long lines of shafting have not been already introduced to furnish power at rates more reasonable than it can be obtained for at present.

The inability of motive power to do its work, or its temporary failure through accidents, is a serious loss, since repairs are equally costly and dilatory. To the best of my knowledge, there is not a shop in the oil region where a boiler and steam-engine can be renovated, if badly broken down. The demands upon the shops at Buffalo, Pittsburgh, Philadelphia, Paterson, and Newburgh, have hitherto been such that new engines could be obtained only after months of delay, as well as by paying double the prices which they would formerly have commanded. Why, among the thousands of manufacturers and mechanics who have visited Venango county, the establishment of a boiler-shop and engine-shop should have been overlooked, I am unable to conjecture, unless it has been owing to the extravagant prices charged for land and services of all kinds, the cupidity of speculators thus checking the growth of industry, the settlement of the country, and in the end cutting its own throat.

Getting out of fuel is one of the many temporary annoyances experienced by oil companies. At first sight, this would hardly appear possible, in view of the abundance of fire-wood still left in the country, after the more valuable timber has been cleared off and sawn into lumber. Petrolia has for boundaries on two sides the coal-fields of Pennsylvania, the beds cropping out on the hill-tops, both toward the east and the south. In both directions the seams are of sufficient thickness to render mining profitable in ordinary times. But the insane spirit of speculation has caused coal-lands to advance in price from one hundred dollars to two thousand dollars per acre, and thus practically shut up one of the chief sources of wealth in that region. The people have to depend upon the irregular and uncertain supplies received from Pittsburgh by steamer, paying quadruple prices therefor. With railroads and steamboats utterly insufficient to transact the business thrown upon them, and no common roads deserving of the name, it came to pass that bituminous coal commanded from sixty to eighty cents per bushel in Oil City, during the past winter; while along Oil Creek it advanced to one dollar and even one dollar and twenty-five cents per bushel. Here again is seen the consequence of that shortsighted selfishness, which, refusing to take thought for others as well as for the morrow, leaves the highways (a sad misnomer!) to take care of themselves. Horse railroads along the principal bottoms could not fail to be exceedingly remunerative.

Sometimes I have felt indignant at beholding a whole *posse* of men loitering idly about a well, because they were out of coal, which might not arrive within a full week. There lay along the river-margin, within a quarter of a mile, drift-wood sufficient to supply fuel for a fortnight; there was abundance on the slopes immediately above, to be had for a trifle, in addition to the labor of chopping and rolling it down. The officer in charge was too consequential to step so far out of his beat; the engineers and others were too indolent to work when they could do without it. Between the two, wells produced after a fashion, and shareholders got dividends—in faith, hope, and charity!

In the chapter on statistics will be found occasional references to *the blower*, a simple apparatus, first introduced into the Sherman well, and since then tried in several others with various success. The Sherman had flowed very largely for a time, but given out; resort was then had to the pump, which also at length proved insufficient. Finally, the manager introduced the instrument mentioned with quite satisfactory results, the yield coming up to between forty and sixty barrels a day. The blower consists simply of an iron tube, one inch in diameter, which is let down into the well outside of the fixed tube, the lower end of the blower being bent round, so as to pass upward into the orifice of the other. A column of air is then forced down the small pipe, from which it passes into the larger, and assists the natural action of the gas in forcing up the petroleum. By virtue of this simple contrivance to co-operate with Nature, that well has gone forward for some months without any material diminution that I could learn. In other cases, however, the result was far from being equally satisfactory. The manager of the Noble and Delamater well tried the blower, under which the total yield amounted to twenty-five barrels, when the oil ceased to flow. In others, again, after using the blower without success, it has been taken out, and replaced by the pump, with a decided improvement in the yield. It deserves mention that these instruments are not all similar in design; and that some are regarded worthless under all circumstances. Before purchasing, it would be well to obtain a guarantee of satisfactory results, provided it can be proved that oil exists in the well to be subjected to the experiment. On this matter, as of every other, the experience of both practical and disinterested men is highly desirable.

The cost of transporting petroleum, in bulk or barrels, to the nearest point of shipment, is one of those obstacles now felt in preventing the further extension of that great enterprise. As a specimen of the rates prevailing since the opening of navigation, I may state that boatmen are in the habit of charging one dollar per barrel for conveying it from Petroleum Centre four miles up to Shaffer's station, or eight miles down to Oil City. Before the opening of the railroad from Franklin to Oil City, it was customary to charge two dollars per barrel for transporting it seven miles, during the winter season. A railroad between Shaffer's and Oil City could make money by carrying a barrel the whole distance for twenty-five cents.

But in spite of this obvious truth, staring stockholders and managers in the face—in spite of troublesome delays as well as extravagant changes, the Petrolians are content to trudge through mud and grease, the noblest of God's subordinate creatures being murdered every year by thousands. I know of no better means for bringing the ferocious drivers to something like feeling than to apply to *them* the blows they are wont to administer to the broken-down quadrupeds, often in the agonies of death. In one instance at least, an earnest threat from a spectator to give blow for blow had the desired effect, and the poor creature was permitted to heave his expiring groans free from the devilish treatment which he was then perhaps too unconscious to feel.

The newly imposed tax of one dollar per barrel laid by the general government on crude oil, is a subject of much complaint, not only on account of its pecuniary weight, but for the other annoyances in connection with it. One of these will doubtless be the frequent exercise of that rule known as "reduction descending," in regard to the reported yield of many wells! But this topic belongs rather to the financial than the mechanical head, and I pass it by for the present with merely alluding to it.

The losses and troubles arising from owners or agents withdrawing the tubes and leaving idle wells open, have been noticed in a previous chapter. Elsewhere the necessity of stringent legislation is urged upon the state authorities.

No inconsiderable loss was occasioned last winter from the inability of the great railroad companies to transport the petroleum (whether crude or refined) eastward. The consequence of this was an immense accumulation of it at various points, where it could neither be shipped nor made a means of raising money, however much the managers might be in want of it. Dividends, of course, were out of the question. The industrial changes consequent on a restoration of peace will probably prevent a repetition of this misfortune for some years.

A decided drawback on the business as a whole, is *the great fluctuation in prices* happening every little while. In March last, the price of crude illuminating oil was from ten to twelve dollars per barrel at the wells; at one time in April it had sunk to three dollars, a fall of seventy per cent! Even this *lapsus* was nothing like so ruinous as that which was experienced in 1861, when the great flowing wells down Oil Creek were first struck. For many weeks petroleum was sold for twenty-five and even ten cents per barrel—in fact, it was not thought to be worth barrelling at all. There was no resource but to wait and extend the introduction of this article, which was duly accomplished in time, though not till tens of thousands of barrels of it had gone to waste. The wild excitement of last winter and spring, attended with such a rush of people to the oil regions, had its influence in lowering prices last April. If most of the wells were really as successful as they are represented to be, there can be no doubt that the price of oil would sink to a still lower figure than three dollars per barrel.

The high wages paid, whether directly for labor or indirectly for its products, have been in part attributed to the war; with its close there cannot fail to be a material decline in both. It is questionable, however, whether companies have been as much losers from the high standard ruling as from the unsteadiness of a large body of the workmen, at a time when employment is abundant and well remunerated. This will also come to a close.

Last, but not least, among the troubles to be expected, is "*the chapter of accidents*," arising from flood, fire, etc. The mere mention of this suggests what a story of losses and crosses has been told since the end of last March, when, in consequence of the melting of a whole winter's accumulations of snow, the runs, creeks, and rivers overflowed their banks and inundated the lowlands, piers, wharves, etc., sweeping off property estimated by some as being worth five millions of dollars! The damage done directly to wells, together with the consequent falling off in their production, I have no doubt will have amounted to as large a sum as the direct destruction of oil and other property. One-fifth of the immense aggregate, applied to the construction of roads and levees, drains and fences, with reasonable sagacity in forecasting the advent of such a visitation, might have warded off the effects of that frightful calamity; as it is, companies and individuals must pay the penalty of establishing conditions in which the lowest form of selfishness is the only recognized principle of action. To this hour Petrolia has refused to profit by that calamitous dispensation. It has opened no water-courses; thrown up no embankments to be used for highways as well as levees; has raised neither well nor tank one foot above its former level; but, with the placid unconcern of a believer in fate, resigned itself to the next flood that may desolate its valleys and inundate its streets. The influx of capital has been so unprecedented, that some of its people may have imagined they can snap their fingers at the natural laws; but these will, in the end, assuredly vindicate themselves.

In still another mode do losses by flood occur. During the summer months, Oil Creek shrinks to a mere brook, insufficient to float even a skiff in many places. At this season it has been customary to create artificial floods or freshets by the construction of dams near the head-waters. On a given day in the week, these are opened and the pent-up waters let flow. The scene is apt to be highly exciting and withal not free from danger, as the newly-emancipated floods rush madly down the valley, bearing scores of huge flat-bottomed boats, all heavily laden with the products of the wells. Sometimes boats, laden in bulk, collide with those laden with barrels, and both again crash against others which may be moored to the margin of the creek, breaking them into fragments and sending their cargoes adrift. It is customary to give notice in advance of these visitations; but with all the precautions adopted, they are apt to be attended with more or less damage.

In speaking of *the effects of fire*, it is gratifying to find that the conflagrations which broke out with consequences so tremendous in the early history of this enterprise, have not happened in vain. Some of these calamities were so appalling that they might almost be termed *national* evils. In one of them, near Rouseville, twenty-seven persons lost their lives; and seldom, indeed, has a well taken fire without more or less destruction of human life, as well as immense amounts of property. It requires very little stretch of the imagination to comprehend the effects of a single spark falling in the wrong place when a stiff breeze is blowing. With hundreds of derricks, tanks, engine-houses, etc., on a single bottom—all of wood, and so saturated with oil that they are as inflammable as tinder—nay, with streams and pools of petroleum covering large portions of the surface, it requires but a spark to cause the whole magazine to explode. It has even sometimes happened that upon striking oil suddenly, this liquid would spirt upward, fall on the engine-house, and take fire from the furnace, enveloping all the neighboring locality in flames. Accidents of this kind have taken place so often from laborers or strangers smoking near the wells, that the practice is prohibited, and the stranger finds numerous cautions to that effect wherever he goes.

"When the burning well," writes a correspondent, "happened to be near the creek, which was frequently the case, a new danger threatened. The boats, loaded with oil, took fire, and burning their lines, went adrift down the stream. As they passed the tank-boats, filled with oil in bulk, the flames spread to them with the rapidity of lightning; a single flash and the whole boat was in flames. The burning boats continued in their course of devastation, setting fire to every thing they touched on their route. The bursting of tanks covered the stream with oil, which took fire and added to the terrible grandeur of the scene. A great fire of this kind occurred more than a year ago, when blazing boats came down the creek, and, plunging among a large fleet of loaded boats at Oil City, set them on fire; and the whole blazing mass swept down the Alleghany, burning the Franklin bridge as they passed, and spreading terror for miles along the river. . . . To extinguish a burning well, the only means discovered is to play on it with steam from the boiler of a neighboring engine, or to heap the mouth of the hole with earth. To accomplish the latter is a work of great difficulty and danger, for the heat thrown out by burning petroleum is intense. For this reason, a flowing well that takes fire frequently burns for many days before the fire can be extinguished."

The wrecks left by such a visitation are scarcely noticeable, consisting of a few long hoops that have fallen from the tanks. Every thing else is sure to be licked up. Last winter, several works, about a mile above Petroleum Centre, took fire and disappeared; and not a few of the derricks arising in different parts of the oil regions, supposed by the visitor to indicate the sinking of new wells, are simply replacing old works which have perished by flood or flame.

In perusing the various accounts of conflagrations and their causes, it may occur to some, that one source of danger is the escaping gas becoming a conductor of electricity during thunder-storms, as the exhalations arising from trees and barns filled with grain are known to be. Numerous inquiries, however, have led me to believe that the gas from wells does not act as a conductor, as no fire is known to have thus originated. Otherwise it would hardly seem possible to escape a sweeping calamity nearly every week during the summer season.

But whether or not delays or disasters be experienced from flood or flame, the lack of fuel, defects in engines, disorganization in labor, obstructions in the wells or the tubes, the chilling blasts of winter, or the descent of surface-water, it is an established fact, that the springs themselves decay and finally cease to yield petroleum in paying quantities. They may be nursed, and humored, and coaxed, and petted, until it would seem that every imaginable whim was gratified. But no mechanical or chemical application can *easily* extract sunbeams from cucumbers, or oil from an empty well. For that the real trouble with most of the abandoned concerns is *exhaustion*, does not admit of a doubt, all the asseverations of owners, managers, agents, mercenary writers, etc., to the contrary notwithstanding. The period after which tired Nature will say "Enough," and refuse to contribute another quart, may last three months or three years; but it will assuredly come to an end. It may be prolonged for a time by various schemes and appliances; but animal life does not more certainly, in the long run, yield obedience to the natural law of decline, decay, and dissolution, than must the oil-spring. For nearly four years, the Economy wells have resisted the fate which befell all others of the same age; but even they are producing much less than they did in 1861, and one has become quite feeble. The others, in time, will certainly follow, especially if numbers of new works are to be put into operation close by. Let all who mean to invest their money in oil stocks "stick a pin there."

CHAPTER VI.

STATISTICS OF PRODUCTION.

HAVING described the physical features of Petrolia, its present appearance, "the manners and customs" of its inhabitants, the mode of drilling wells, and the obstacles to be overcome afterward, the next point in order, and that for which this little volume will be most eagerly perused by many, is to give statistics of the present and past productions of the wells, as ascertained by careful inquiry upon personal examination. The collecting of these consumed fully three weeks. Sometimes a mile or two constituted the length of my cord for a whole day. I made it a point to visit every well in operation, and to compare the yield reported by workmen with that given by officers and neighbors, examining the discharge, and suggesting doubts or inquiries, as the case might be. In order to get the *whole* truth, it was sometimes necessary to inform the men that I was *not* a government agent; at other times, to probe statements very closely and take additional testimony. By the use of methods best known to a newspaper expert, I think I arrived as near the truth as it was possible for a stranger to do in any thing like a reasonable time. As to fidelity in reporting, I am not conscious of having misrepresented a fact, a figure, an impression, or even having *colored* it in the slightest degree, with a view to benefiting or running down any person or interest. Opportunities and even tempting offers to assist in finding purchasers for wells were not wanting, whatever may have been the motive in rejecting them. It was with a deep conviction on my mind that the subject had never been investigated as it ought to be; that the hard financial and industrial facts had not been collected by laborious, judicious, conscientious men, with a view to publication, that I felt it my duty to take hold of it, in spite of physical obstacles and the predictions of many who pronounced the undertaking an impossibility.

In order to present the greatest amount of matter in the least space, and in something like a systematic form, I have divided the country into sections or farms, lying contiguous to each other, and having points of resemblance. The first of these is the well-known

WATSON FLATS.

These consist of several hundred acres of low and nearly level land, lying between or on each side of the two principal branches of Oil Creek. It has some appearance of having been at one time the bottom of a lake, before the creek, lower down, reached its present depth. At some points on it, as well as further down the creek, pits were found, from five to eight feet in diameter, and from fifteen to twenty feet deep, their sides protected by "wooden walls," which had been carefully joined. The fact that trees, several feet in diameter, have been found growing immediately above those pits, proves that they must have been sunk before the white race made their appearance; and the inference is that the western mound-builders operated in Petrolia perhaps thousands of years since. It is said that the Cornplanter Indians have still traditions to that effect.

The wells on that flat have some characteristics in common. None of them yield oil by *flowing*. The production of the most profitable does not exceed sixty barrels per day. On the other hand, very few wells have been sunk in that locality that did not pay expenses. It is believed there has not been *one* that did not contribute more or less. This "territory" may, therefore, be set down as perhaps the safest in all Petrolia, but the one offering fewest rewards in the shape of rare and dazzling prizes. Deep well, sunk to more than twelve hundred feet and still in progress, is on this tract.

The present yield of the active wells there may be set down as between the figures underneath, both having been given by persons on the ground, and one (the lower) by those who professed to know, and I believe did know, of what they affirmed.

U. S. Grant Well.—Sunk four hundred and sixty feet. Engine driven by gas from the well. Discharges a constant stream of oil and water, but increasing and decreasing every few seconds. Yields from thirty-five to forty-five barrels every twenty-four hours; probable average, not far from forty.

Ingersoll Well.—Owned by New-York and Oil Creek Petroleum Company. Believed to be the best on the flat. Pumps a steady stream of between fifty and sixty barrels daily. Had been yielding only three weeks at time of visit. They pumped all winter, and got nothing but water for months. Has yielded as much as seventy barrels a day. Depth, four hundred and fifty-eight feet.

Old Abe Well.—Was opened in April last, a week before visit. Yields thirty barrels every twenty-four hours. Depth, two hundred and sixty feet.

Kellogg Well.—Yields from ten to fifteen barrels per day. Depth, four hundred and sixty-two feet. Was opened last February. Discharges by fits and starts only.

Continental Well.—Has broken down two or three times lately. Was idle, when visited, being re-tubed. Famous in the past.

Katz Well.—Was once the best on this flat; now gives little except water. Tools fast in the bottom, which is assigned as the probable cause of failure. Lay idle from August till lately.

Nellie Binninger Well.—Has been in operation about three months. Very irregular in yield, which ranges from one to fifteen barrels per day.

New Well.—Owned by Pennsylvania Oil Creek Company. Prospects, when finished, considered good.

Watson and Childs's Well.—Is old, but recently renovated. Used to flow sixty barrels per day. Has been idle for nearly three years. Cannot yet say what she is likely to do.

Baker Well.—Yields about twenty-two barrels per day. At one time, gave as much as forty, and four years ago flowed one hundred and sixty barrels daily. Uses her own gas for fuel.

Ostrom Well.—Four years old. Yields about thirty barrels per day. Depth, four hundred and twenty feet.

Elliott's Well.—Was yielding forty barrels daily till the spring freshet, which damaged the works and suspended operations till visit.

Dunn Well.—Average yield given at forty barrels per

day for four weeks. Little discharge of oil when visited, it being late in the day. Has regular pulsations or tides, discharging oil freely in the forenoon and slacking up afterward.

MEM.—On the Kingsland flat, immediately below Watson flat, are perhaps a dozen derricks, new and old; one of these standing over the famous well sunk by Colonel Drake in 1859. No active operations are visible on any portion of that property.

Dean and Firth's Well.—Belongs to the Bunker Hill Oil Company. Average yield, twelve barrels per day. Depth, five hundred and twenty feet. Doing nothing the day of visit. Opened last October.

Funk Well.—Belongs to owner of that name. Yields ten to twelve barrels per day. Opened last autumn.

May Queen Well.—Belongs to same proprietor, and gives from seven to ten barrels daily. Was also put down last fall, the work done in twenty-five days, to about the usual depth. Believed to be "the quickest time on record."

MEM.—A lot of forty by two hundred feet, close by, has changed hands for the sum of three thousand dollars; purchaser to take his chance. No wells on property, and no royalty on oil, if any reached. Is about one mile from Titusville, and near margin of the "oil-diggings." Learn also that thirty-seven acres, at a short distance, were lately sold for two hundred and fifty thousand dollars in fee simple, the agent of an Eastern company having paid down ten thousand dollars as earnest. Would be glad he had the "eighteen pence within his purse" again; but seller refuses to give it up. A well in that neighborhood, represented as giving fifteen barrels per day, sold last winter for forty-five thousand dollars. It is not paying expenses to-day, and at the time of sale did not give thirty per cent of the quantity stated!

Hinckley Well.—Was never properly tested, (a suspicious phrase.) Pumped thirteen barrels oil one day some time since. Repairing the machinery, and mean to deepen the well. Depth, four hundred and eighty feet. Used a blower, which has been taken out as unserviceable. Belongs to the Titus estate.

Burtiss Well.—Yielded seven to eight barrels per day last year. Idle in winter, but starting again. Said to be "good show," and there is certainly *water.* Depth, four hundred and eighty-five feet.

Cap Well, or No. Seven Parker Petroleum Company, has not been running steadily for some time. They are pumping and expect to get oil. Depth, six hundred and fourteen feet.

Eckhart Well.—Just starting. A little oil on the brine, though barely one barrel per day. Depth, six hundred and thirteen feet.

Palmer Well.—Has been lying still for some time. Gave from twelve to eighteen barrels per day formerly. Making preparations to resume work. Depth, two hundred and forty feet. The *Keystone,* close by, same tale as Palmer.

Utica Rock-Oil Company's Well, No. Thirty-nine.—Opened a few days before visit. In two days gave twenty-eight barrels, and yield increasing. Depth, four hundred and fifty-seven feet.

Well No. Three.—Old concern reöpening. In three weeks gave two or three barrels. Depth, six hundred and fifty feet. They say the tubing is out of order. These wells are the highest up on the east branch of Oil Creek of works in actual operation. Distance from Titusville, one and a half miles.

Tripler Well.—About four years old. Gives twelve barrels per day. Depth, about five hundred feet.

Old Hinkle Well.—Flowed one hundred barrels per day at first. Subsequently abandoned. Was pumped some weeks, at one time giving one hundred and twenty-nine barrels in three days. Owners did not consider this profitable, with petroleum at twenty-five cents per barrel, and allowed well to rest. Preparing to pump it once more.

Glendale Oil Company's Well.—Averages about three and a half barrels per day. Goes by fits and starts only. Not properly tested since deepening.

Binninger and Sanger Wells.—One is not pumped. Other is six hundred feet, and newly tested. Gives between twenty and thirty barrels per day, on average. An outside account says between twelve and fifteen.

Shermouth Well.—Old and abandoned. At one time good for forty to fifty barrels per day. Sold to new company, late owners guaranteeing an average yield of thirty barrels.

No. Two Parker Well.—Pumps about four barrels in ten hours. Two years old. Depth, five hundred feet.

Abbott Wells.—Two in number. Pumping small quantities of oil, but the exact quantity could not be ascertained.

Above is a list of all the wells in actual operation, or having been such within a few days of my visit. The number of idle or abandoned works I estimate at fifty, within two miles of Titusville; but in a goodly number of these preparations are in progress for resuming operations. The effects of the freshet are still painfully visible. The number of new derricks within the same distance is fully one hundred — perhaps one hundred and twenty-five; but at a considerable number of these there are no signs of activity.

After sinking to the depth of four hundred and fifty feet on the Watson flats, without finding oil, I have very little faith in further deepening. Most of the experiments made in that direction have been unsuccessful, and none have amounted to much.

Wells on the west side of the principal fork, on the whole, pay better than those on the opposite side. The ones given at the head of this list are all on the west side. I allude to this only to dispel an error prevalent in some quarters, that the east side of all streams is more productive than the west—a vagary refuted by understanding the physical formation of the country. Wells along the base of the heights north-east of Titusville pay moderately, their average being from eight to ten barrels per day. On the opposite slope they have not yet advanced sufficiently near completion to afford information as to the extent of the oil-field in that direction. Slopes, as such, have no necessary connection with oil-bearing rocks.

For nearly four miles below the Watson flats little progress has recently been made, the few wells sunk here and there appearing to have been unprofitable. Still, other persons are about to make fresh experiments, and a number of new wells are going down. The next point of interest is

MILLER'S FARM,

six miles below Titusville, with a railroad station and a bridge across the creek. Below is a list of the productive wells:

New-England Well.—Belongs to company of same name. Opened about five months ago. Yields from twenty-five to thirty barrels per day. Average well up to thirty. Was the only one then in operation at time of visit, the others undergoing repairs. Is on east side of the creek.

Hemlock Well.—Is on west side. Pumped about eight barrels per day through winter; but it and the *Kerosene* were undergoing repairs. Hemlock had been going about one year. Kerosene yielded about twelve barrels per day.

There are a few other old wells on this farm, chiefly on the east side; but all appear to have given out. About ten new derricks erected, at half of which work is actually going on. A small refinery at the station.

At Shaffer's, the railroad terminus, out of a dozen old works close by there was not one yielding oil at the time of visit. Great expectations from some new ones on opposite side of the creek. Number of old and idle wells from Watson flats to this place, probably fifty, with nearly as many new ones in progress. Much activity at Shaffer's in unloading oil from the boats, and shipping merchandise from the railroad. An "excellent show" of portable engines—say a quarter of an acre at least.

For the next mile and a half a masterly inactivity prevails along the widening bottom. On that space, I think, are one hundred idle derricks. Disinterested men pronounce the "territory dry;" and the solitude of the scene cannot be explained by freshet or other temporary breakdown. At length we find, on

THE FOSTER FARM,

Vanita Well, No. Two.—Belongs to the Indian Oil Company of Philadelphia. Flowed about three hundred barrels per day, after opening, last fall; but has since fallen off to about four barrels.

Porter Well.—Belongs to same interest. Flowed from May till September, 1864, at the rate of one hundred and fifty barrels per day. In November the pump was applied, and well now yields about ten barrels, supplying gas to run the engine.

Zinc Well.—Belongs to Irwin Oil Company of Philadelphia. Yields no oil worth mentioning. Workmen expect it to produce after exhausting the salt water. Began operations, after lying idle, a short time since.

THE McELHENNY FARMS.

Some of the most famous wells in the oil region were struck, in 1861, about two miles below Shaffer's, on or near the McElhenny farms. The flats there have expanded to about two hundred and fifty yards in width, and are forested with old and new derricks. The principal works are (or were) the following:

Noble and Delamater Well.—Is on the east side. Opened in 1862. Flowed from eighteen hundred to two thousand barrels per day for six months, when it began to fall off. In spite of low prices, partly caused by its success, made its owners (two poor men) millionaires. About the middle of last winter ceased to flow, when a blower was put in. The trial resulted in a total yield of about twenty-five barrels. Had been idle for some time before visit. No symptoms of renewed activity.

6

McKinney Well.—Same side of the creek. Yielded about fifty barrels a day by pumping, a short time ago, but broke down for a time. Gives at present about thirty barrels.

Craft Well.—On Bull Run, entering creek on east side. Opened about the first of April, but a "fool" of the right kind, as it flows two hundred barrels per day, according to owners' estimate; and I think the yield well up to that figure. No falling off at time of visit. A steady stream from the tube, with little fuss and fury.

Railroad or Boughton Well.—On same stream. At first pumped seventy-five barrels per day. Last November, valves got out of order, and well has not since been in operation. Workmen very costive in regard to information. One was quite rude. A decidedly unfavorable impression about the concern.

Caldwell Well.—Just testing. Fair prospects of success. Is situated above and close by the Noble well.

Caldwell Well, No. Four.—Lately tested. Tools fast in bottom. Prospect indifferent. Another well on lot No. Five. Opened two weeks and pumping water.

Irwin Well.—On west side. Pumping from twenty to twenty-five barrels per day.

Crocker Well.—Noted old concern in its day; but has yielded little except water the past winter. A dispute with neighboring proprietor, who had tapped the same vein. Late Crocker owners say they offered fair terms, which other party rejected. They then withdrew their tubes, when, *presto!* the rival was obliged to purchase Crocker or go without any thing. A controversy not uncommon in Petrolia.

Lon Well.—Near old Sherman. Just testing at time

of visit, and good for ten to fifteen barrels, it was said, per day.

Sherman Well.—One of the historical landmarks of the oil regions. Was one of those which inaugurated the revolution of 1861, causing scores of wells to be abandoned, through reduction in price. First yield, fifteen hundred barrels per day, by flowing; then fell off to six hundred; dropped down to one hundred; and finally stopped altogether last fall. Tried the pump for a time; but again became unproductive. Next, the proprietor put in a blower, which brought up an average yield of fifty barrels per day, sometimes sixty barrels. An outsider, who pretends to know, reports the amount at forty barrels. Mr. Sherman was originally a poor, but very energetic man. He is still energetic, but not poor. Had to borrow means enough to finish the drilling; is now one of the magnates of that country.

Fertig Well.—Started with sixty barrels per day, which gradually receded to ten, and then recovered somewhat. Now ranges between ten and twenty barrels, the average being probably fifteen. Is nearly two years old. Greatest decline during the cold weather, on the approach of which yield fell off one-half.

MEM.—The proportion of productive wells on this farm and for some distance below, (say one mile beyond the Sherman well,) does not exceed one to every seven or eight of the whole number sunk. In some cases they are endeavoring to resuscitate the idle concerns. At one place counted more than twenty derricks, the machinery all standing still. Beyond those were ten or twelve belonging to a Baltimore company, having been idle for some months; but a portion of them would probably be tried anew before long. Continuing the journey downward, the next productive work is the

Buckeye Well.—Yields from eight to ten barrels per day. Has just been tested anew, after being renovated. Put down about two and a half years ago. Depth, five hundred and fifty feet.

Briggs Wells.—Four in number—the Briggs, Sabine, Forest City, and Cayuga. First-named does not pump steadily. Betimes gives nothing, betimes as much as seventy barrels per day. Ordinarily ranges from twenty to twenty-five. *Forest City* yields about forty-five barrels per day. Was opened in 1863, when flowed one hundred and fifty barrels. Proprietors thought that was "not much of a shower," and let it go to the creek. Started again after oil had risen in price. *Cayuga* yields about eight barrels. Last summer gave thirty, but fell off in winter. *Sabine* in operation about two months, and when visited gave thirty-five barrels per day. All these wells are pumped.

Huyde Koper Well.—Uses both a pump and a blower, which bring up forty barrels per day. Has been in operation two or three years. Depth, four hundred and sixty feet.

Mount Vernon Oil Company's Wells.—Four in number. Two are doing little or nothing. One is said to average forty barrels per day. The other somewhat out of order and not going at the time.

THE FUNK AND BOYD FARMS.

Village named Funkville, and already described. Former owner, one "Captain" Funk, who died recently one of the princes of Petrolia. The first flowing well ever struck was on this property; and one of the most famous in the country is still productive, namely, the

Empire Well, No. One.—Sunk in the summer of 1861; and for a time flowed over two thousand barrels per day, the yield gradually declining, and then dying out. Last year a blower was put in with moderate success; but with the pump has done better. On some days yields nearly one hundred barrels; but the average is between sixty and seventy. Flow steady.

Empire No. Two.—Idle, her tools having stuck fast. Never did any thing. *No. Three* is about eighteen months old, and yields fifteen barrels daily, with no falling off perceptible. At first was a flowing well.

Oceanic Oil Company's Wells.—Four in number. *No. Three* was testing with good prospects at time of visit. The others were not going.

Jenkins Well.—Has been flowing for two months. Average yield given at one hundred barrels per day, with no reports outside to the contrary.

Buckeye Well.—Ranges from twenty to forty barrels per day, with an average yield of nearly thirty. Flowed at one time six hundred barrels daily, sending up a column of oil, when struck, described as being one hundred and fifty feet high. Now pumping.

Hibbard Well No. Two.—Used to flow over one hundred barrels per day. Now pumps about twenty. Two years old. Depth, four hundred and eighty feet.

T——n Well.—Pumps about sixty-five barrels. Uses its own gas as fuel. Been in operation about two months. One-half interest belongs to Oceanic Oil Company; the rest to Mr. Sherman. (N. B.—The name is blotted in my note-book.)

Hatch Wells.—Both idle, but preparing to resume work. One ordinarily yields about fifty barrels per day, and is now flowing nearly that quantity. Have set it down as good for forty. Another gives only two or three barrels oil, but gas enough to supply eight engines.

Hadding No. Two Well.—Has been in operation three years. Does not go steadily. When in order, yields twenty barrels per day. Estimate its average at ten to twelve.

American Well.—Has been in operation only a few weeks. Average yield reported at thirty barrels per day. Water not yet exhausted.

Densmore Well No. Eleven.—New pumping well. Yield uncertain; but supposed to be sixty barrels a day.

Well No. Five.—Also new and pumping. Yield rated at seventy-five barrels per day. Two other wells doing nothing. Belong to same interest as last.

Olmstead Wells.—*No. Three* produces, by pumping, fifty-five barrels per day. Is eighteen months old. *No. One* has been flowing at rate of seventy-five barrels for a day or two. *No. Four* pumps about seventy barrels daily. *No. Two* has been idle some years.

MEM.—Very few wells on this flat reach to what is called "the third sand-rock," their average depth being between four hundred and fifty and four hundred and seventy-five feet.

On the Boyd farm counted about forty derricks, nearly all idle. Most of them on east side of creek. A few were approaching completion. On the whole, this farm, like the lower portion of the Funk flat, has not been very productive. Fire broke out during the winter, and did serious damage. Near the lower end are—

Wood and Wright's Well.—Pumping about three barrels per day, when in operation, but now idle; also a well, said to belong to Wood & Company, and reported to pump, "by head," seven or eight barrels.

BENNEHOOF RUN

is the name of a streamlet, perhaps ten miles long, discharging into Oil Creek from the west. It has lately obtained some celebrity from a flowing well struck on it and named—

Lady Herman Well.—Belongs to a man named Herman, who had the good sense to name it after his wife—so it is said. Average yield, between sixty and seventy barrels per day. Oil struck about two months since.

Mowbray Well.—Old and abandoned for a time. Cleaning out and testing anew. Used to be a flowing well. New owner at work.

Old Warren Well, No. Two.—When in operation, yields about ten barrels per day. Been re-tubed and is testing.

Mem.—At different times numbers of wells have been sunk on this run. Mostly now idle. Number of old works, about fifteen. At least as many new derricks have gone up or are going. Preparations for a vigorous campaign quite active here.

THE WASH. M'CLINTOCK FARM, OR PETROLEUM CENTRE.

The principal wells on this tract are on the west side of the creek, and altogether the place is one of the most prosperous in the rural districts of Petrolia. Half a dozen refineries are on the farm, and great activity prevails in sinking wells, building houses, etc. Most of the flat belongs to the Central Petroleum Oil Company, who lease out half-acre lots for a royalty of one-half the oil and a bonus of ten thousand dollars in money. In the incipient village, lots twenty by one hundred feet lease for one hundred dollars per annum, all improvements reverting to the company at the end of five years; while they have the right of entering upon the premises at any time, after thirty days' notice, to sink a well. Even on these steep conditions leases have been sold for a considerable premium!

There is something unusual in the formation of this bottom. At one time the creek made a wide detour to the westward, describing a horse-shoe course for half a mile, and inclosing an island two hundred and fifty yards in length. More recently it has forsaken that part of its bed altogether and confined itself to the eastern and shorter channel, leaving the island a ridgy hill, about seventy-five feet high. The most productive wells are on the bottom, which was formerly the bed of the creek. That stream is crossed by a substantial bridge, the only one between Shaffer's and Oil City. Situated four miles below the former and eight above the latter, the carrying trade divides near Petroleum Centre, going upward or downward by boat, according to circumstances. There is a hotel in the village. Subjoined are the wells in operation:

Central Petroleum Company's Well, No. One.—Has been flowing for ten months. Gives twenty barrels per day on the average. Flows spasmodically, stopping altogether for forty minutes, then coughing, and expectorating, and discharging for five or six minutes. These reports and eructations increase till they sound like the exhausts of a steam-engine, and can be heard for quite a long distance. This fit having spent itself, the discharge gradually subsides.

No. Ten Well.—Belongs to same company. While drilling, about four weeks before visit, oil spirted forth, striking the top of the derrick and driving off the workmen. It flowed in a steady and powerful stream toward the creek. Many hundred barrels lost before the well could be tubed and the product saved. Is now calm and beginning to yield. Depth, five hundred and fifty feet. I rather guess than estimate the product at twenty barrels per day.

Bluff Well.—Belongs to same company, and gives, by pumping, four to five barrels per day. For some time, flowed from five hundred to eight hundred barrels daily; but this gave out altogether.

Coldwater No. Seven Well.—Pumps twenty barrels per day. Occasionally flows a little. Was sunk last autumn. Depth, five hundred and eighty feet.

Coldwater No. Three Well.—Newly opened and flowing at the rate of about twenty barrels per day. Struck the vein of *No. Four*, which has since gone dry or pumps only water.

Coldwater Wells, One and Two.—Stopped, waiting for engines. Used to flow about twenty barrels per day each. *No. Five* pumps ten barrels a day. *No. Six* in progress.

Central Petroleum Company's Well, No. Two.—Pumps about fifteen barrels daily. *No. Three* flows and pumps about twenty barrels. *No. Four* idle. *No. Five* flows about sixty barrels. *No. Six* suspended. *No. Seven* pumps about fifteen barrels. *Nos. Eight and Nine* in progress.

Fox Well, or No. Four.—Flowed one hundred and fifty barrels daily at one time. Now produces about forty. Keeping up steadily to that figure. Belongs to same interest.

Swamp Angel Well.—About fourteen months old. Flows about fifteen barrels per day.

Fowler and Custer's Wells.—Two in number. Produce about twenty barrels per day. Leased from Central Petroleum Company for half the product.

At no other point on the creek is the proportion of idle or unprofitable wells smaller than on this farm. On the upper part of it, however, a number were doing nothing at the time of visit; and I estimate the number of paying wells at this moment at one-half of those which have been put down. It will be seen that of the productive works, a very considerable number were sunk the past year. On both sides of the creek numbers of new works are in progress, climbing the hill-sides to the height of one hundred and fifty feet.

THE EGBERT FARM

lies nearly opposite to Petroleum Centre. It is sometimes known as the Hyde and Egbert, having been owned by two persons, the latter a country physician, and both in moderate circumstances. It is now the most productive portion of Petrolia, with the possible exception of a farm on Cherry Run. It is also near the geographical centre of the Pennsylvania oil region. Its principal well, and the most profitable in the country at present, is the—

Coquette Well.—Was opened last October, when took to flowing six hundred barrels per day. No actual test of its production has lately been made. Superintendent reports the last at five hundred and twenty barrels per day. Other reports place the actual amount at four hundred to four hundred and fifty barrels. Well has a long double range of tanks, whose aggregate capacity is fifteen thousand barrels, and connecting, by a set of pipes, with the creek, so that boats can be loaded in bulk. The discharges of this spring occur in regular successions of puffs and jets, noticed elsewhere; the series repeated five or six times per minute. About one-seventh of the interest in it recently sold in Philadelphia for three hundred and fifty thousand dollars. Property mostly owned in that city.

Maple Shade Well.—Struck in August, 1863. Flows about fifty barrels per day. Gave over one thousand barrels at one time—in fact, one of the glories of that country. Is about nineteen months old. Present yield reported to be keeping steady. One year ago, was burned down, together with ten thousand barrels of oil and twenty thousand barrels of tankage. Belongs to Maple Shade Oil Company of Philadelphia, who receive one-sixth of the oil produced on three-fourths of Egbert farm, and five-twelfths of that on remaining fourth.

Gettysburgh Well.—Opened last autumn. Yields from twenty-five to thirty barrels per day, with tools fast in bottom. Flowed at first about ten barrels daily. Working interest leased for one-half oil.

Gimlet Well.—Was yielding from ten to twelve barrels per day before stopping to re-tube. Tried the blower, but had to take it out. Now applying the pump to exhaust water.

A well in operation on the Dalzell farm is said to give thirty barrels daily. Further particulars not ascertained. A safer figure would be twenty barrels.

Jersey Well.—A famous well, struck in the early part of last summer. Yielded three hundred and thirty barrels (flowing) for five months, on single days running up to four hundred barrels. Capacity of tanks, ten thousand barrels. Present yield supposed to be between two hundred and two hundred and fifty barrels daily; but has not been tested lately. MEM.—If the yield were increasing, figures would be duly published. Hence, infer that production is below two hundred barrels. P. S.—I have since learned that the Jersey well gave out in May, the spring having been tapped by another well.

Keystone Well, No. One.—A good well at one time, but now idle.

Kepler Well.—Flowed largely for a time. Now pumps about four barrels per day. One year old.

THE STORY FARM.

The wells on this farm lie chiefly on the west side of the creek. The upper part of this, like the lower part of the Egbert, is rather poor "territory." In that section a bare quarter of the works pays operating expenses. On the best part of the Egbert farm one-half of the wells are idle. Story farm (five hundred acres) belongs to the Columbian Oil Company of Pittsburgh, who have sunk over one hundred wells, of which about fifty are said to have been profitable concerns. Last year they put down thirteen, not one of which paid expenses. The best wells are close to the bluff, at a considerable distance from the creek, and the bottom is nearly abandoned. The company prefer boring on their own account to granting leases; though they have granted some at a royalty of one-half. They have been one of the most successful in the valley. They design to sink a large number of new wells the present season; the number estimated at one hundred. Three refineries on the farm. The principal works are—

Ram-Cat Well.—Opened about fourteen months ago. Original yield, (by flowing,) three hundred barrels per day. Present yield, probably about sixty.

Hubbard Well.—Has lately yielded four barrels a day. Been in operation three or four years.

Flora Wells.—Owned by Western Pennsylvania Oil Company. Two of them give, on the average, fifty barrels a day. One produced over four years, and another, eleven months before stopping. A third flowed one hundred and twenty-five barrels per day for a time, and kept on producing for eighteen months.

Reynolds Well.—Pumped as high as fifty barrels per day for a time; recently gave twenty-five; now pumping out water. Two or three years old, and belongs to Oil Creek and Cherry Run Company.

Columbia Oil Company, No. Two, Well.—Just commencing to yield, and quantity unknown. *No. Thirty-six* about two years old; averages fifty barrels per day.

Titus Well.—Flows ninety barrels per day; at one time gave one hundred and fifty. Is eighteen months old.

Perry Well.—Has averaged one hundred barrels per day (flowing) for fifteen months.

Dorman Well.—Ranges from five to fifteen barrels per day by pump. In operation about two months. Two other wells on same lot; one choked up by mud-slide, the other cleaned and tested with "good show."

Hollow Well.—Has given as high as one hundred barrels per day. Gave lately seventy-five on average. Out of order and undergoing repairs.

Hollow Well, No. Four.—Has been going about four months, and now yields nearly one hundred barrels per day.

Eicholtz Well.—Going two or three years. Ranges

from ten to one hundred barrels per day. Average believed to be from thirty to thirty-five.

Lady's. Well, No. Four.—The only one of five wells on lease now producing oil. Pumps eighty barrels per day regularly. In operation for a year. Started at seventy-five barrels. In winter, averaged only twenty.

Lady's Well, No. One or Old Lady's Well.—Quite famous in the annals of Petrolia. Part of the extraordinary crop of 1861. Flowed five hundred barrels per day at first, and kept on yielding till last August, when gave out. Now idle. Might give one or two barrels, if pumped. Last summer, gave fifty barrels a day, and stopped in consequence of the withdrawal of a set of tubes on adjoining farm. Mr. Wade, the superintendent, it is claimed, was the first to introduce the use of gas in driving the machinery.

Ryder & Co.'s Wells.—One has been running since last August. Yields ten barrels per day. The other just testing, and result unknown. Nothing but water as yet.

Gipner Wells.—Two in number. Average yield of both, eighty-five barrels. One over twelve, the other seven months in operation.

Manor Wells.—Also two, which averaged a little over fifty barrels a day for past year.

Salem Well.—Flowing about ten barrels a day. Supply of gas sufficient to drive two engines. Has been running a year, with no perceptible decrease, whether from use of gas or otherwise.

Painter, Nichols & Co.'s Well.—Pumping about ten barrels per day. Continued for two years at this rate.

Malloney Well.—Idle with tools in bottom. Four years old. Rated as a seventy-five barrel well, but *over*-rated.

Bluff Well.—Complain about bad luck lately. Has been giving from twelve to fifteen barrels per day for about two months. Rather weak in the knees at present.

Lloyd Wells.—Two in number, giving together forty barrels daily. One has been going three years; the other, six months.

——— *Well.*—Re-tubing. Flowed at one time fifty to sixty barrels per day. No reliable statistics lately.

Hamburgh Well.—Flowing between sixty and seventy barrels per day. Gave one hundred at one time. Been nine months in operation.

Old Say Well.—Retubed and testing. Pumped fifteen barrels per day till two weeks ago.

THE TARR FARM

has yielded more oil than any other in Petrolia, or perhaps in the world. A brief notice of the career of its late owner will not be amiss here. "Previous to the petroleum excitement," writes the correspondent of the London *Morning Post*, "the owner (Mr. Tarr) was in great straits, his business of rafting lumber, in addition to the cultivation of his miserable acres, scarcely yielding enough to support himself and family in the humble way in which they lived. . . But the oil adventurers came along, and secured a right to bore, giving half the oil to the landowner. The result was that the Philips well struck oil, and yielded over two thousand barrels daily, which, even at the moderate price then current, yielded a magnificent revenue to the owners and also to Tarr. Other wells were sunk and met with great success, so that the poor lumberman and farmer speedily grew rich. In August, 1863, when the price of petroleum ruled low, Tarr sold half the interest of his farm for one hundred and ten thousand dollars cash, and retired to a handsome residence in the adjoining county. But fortune will not let her favorites alone. The remaining interest in his land and wells increased in value until his daily income was counted by thousands of dollars. During the month of December just passed, he closed out his remaining interest on the creek for two million dollars, at which price he considered himself throwing the property away. Uneducated himself, this oil-prince is spending money lavishly on the education of his family; and some amusing but rather doubtful stories are told of his estimation of the power of money in matters of education." The estate is owned by the Tarr Farm Petroleum Company.

The "developed" portion of this property lies on the east side of the creek, a village of about fifty houses and huts resting on the shoulder of the hill. In some places, the bottom is so thickly covered with derricks, etc., that it hardly seems possible to crowd in another. Tanks are standing together by the acre along the margin of the creek.

Old Philips Well.—One of the oldest on the creek, having been opened in the early part of 1861. Yield by flowing at first very large; one account says fifteen hundred barrels per day. Declined gradually, and at last gave out altogether, when applied the pump. Well now gives about twenty-five barrels per day.

Philips Well, No. Two.—Once the greatest in the world. When opened, in October, 1861, flowed thirty-nine hundred and forty barrels in a single day! A child may play with an elephant; and the flow of this monster could be regulated by a stop-cock whenever the tankage came short.

But trouble came. In December following, the Woodford well was put down within seventy-five feet, and tapped the great Philips spring; the Woodford responding to the figure of nearly two thousand barrels a day. Both wells were a little under five hundred feet deep. The Woodford gave out in December, 1853; but meanwhile it had inflicted serious loss on the Philips, and the lawyers were called in to decide the dispute. It was at last settled by a compromise, and the Philips resumed with a much more sober flow. In 1864, it was shut down for seven months, and in November ceased altogether to flow. When visited in April, it had begun to pump about fifty barrels per day. This well has been productive for two years and a half, and the oil furnished by it would probably suffice to pay the cost of all the wells sunk on that farm.

Monitor Well.—Pumping twenty-five barrels per day, and reported to be gaining. Has gone a week at this rate. Used to be a flowing well.

Smith's Wells, Nos. One and Two.—Just recommenced. Both flowed at one time. *No. Two*, with a blower, now yields about twenty barrels per day; *No. One*, little except water. Both about three years old.

Kirwan Well.—Used to give one hundred and twenty barrels per day. Stopped for the winter, and just started afresh. Before stopping gave eight barrels daily.

Pratt and Sherman Well.—Has done little lately. Yielded over one hundred barrels per day at one time, but now dropsical. Was six weeks in pumping out the water last spring—a too general complaint now on *Tarra farma.*

Sterling Well.—Said to give over fifty barrels per day, but the statement appears very questionable, judging from the discharge. It was largely water, when visited, with not more than thirty barrels of oil.

William Penn Wells.—Two of them are pumping out bilge-water. Owners about putting a blower into the third. The last yields about five barrels daily. Never was of much account.

Woodford Well.—Broke the spell of the Big Philips. Has been idle for three months, with some prospect of yielding oil again.

Barely one-fourth of the wells on Tarr farm are active, and of those it will be seen that several are exhausting the water only. Others were preparing to resume, but it was not thought probable that the brine or brackish water would be pumped out before the middle of May or the first of June. Hardly one in ten continued actively at work during the winter. In consequence of puncturing the ground at every few rods and the withdrawal of tubes, the Tarr farm, from being the best, has become only third-class property. The period during which productive wells now actually yield oil does not average six months in the year.

THE BLOOD FARM

is named after its previous owner, Mr. Blood, now also a petroleum millionaire. The property is owned in part by the Blood Farm Petroleum Association, and in part by the Home Petroleum Company. Property lies on both sides of the creek. Below is a list of the active works:

Lehigh Wells.—Three in number. *No. One* is getting ready. Belongs to Germanic Oil Company. *No. Two* is in operation, and yields about ten barrels per day. *No. Three* gives thirty barrels.

Collins Well.—Information refused. Estimated the yield at not more than ten barrels per day. Four-fifths of discharge is water. Well is reported to give thirty barrels per day sometimes.

Burning Well.—Great flowing concern at one time. Gave six hundred barrels per day at first, but decreased gradually, and is now idle.

Pilkins Well.—Used to give one hundred and fifty barrels a day. Is eighteen months old. Broke down lately, and is not doing much at present.

Barrow and Painter's Well.—*No. Two* gives average yield of twenty to twenty-five barrels per day. New well, just tested. Appearances good.

Duncan Well.—Yields fifty barrels a day; on some days as much as sixty. Has been at work about two years. Owner a heavy operator on Cherry Run and elsewhere.

Ocean Company's Well.—Been in operation only about one week. Gives six or seven barrels per day.

True Boy Well.—Has been flowing about twenty barrels per day for past three weeks; average for past year, about ten barrels. Been running two or three years.

Maple Grove Well.—Pumps about forty barrels per day. Has been going over three years. Began with small tubing, and yielded only twenty-five barrels. With improved apparatus has done better.

Combs and Prince Well.—Been going nearly a year. Yields ten to twelve barrels a day by pumping. Another well on same lease just testing.

Bushnell Well.—Gives between thirty and forty barrels daily. Two or three years old.

Old Reed Well.—Flowed ten barrels per day until the

vein was tapped by another, when the quantity named fell off to a mere dripping, probably not one barrel per day.

Buffalo Well.—Yields about five barrels daily. Discharges, by flowing, about three times per minute.

MEM.—Nearly all the wells on the lower part of Blood farm are unproductive, not more than one in ten showing signs of activity, when passed through. A small number of new ones are going down along the heights. On the upper end, a large majority of them are also idle, having given out or been damaged by flood. Some are deepening, with a view to reaching a lower source than has yet been discovered. Nearly half the derricks are new, but a majority of these are inactive.

CHERRYTREE RUN.

This is the name of a mill-stream discharging into Oil Creek from the west, a short distance above Rouseville. It acquired, last winter, some celebrity from a flowing well known as the

Big Tank Well.—Was yielding over fifty barrels per day, when the works caught fire and burned up. The well has not since been rigged up again. Why it has been suffered to remain so, I have not learned.

Waterson's Well.—Just testing. Yielded by pumping, first twenty-four hours, twelve or fifteen barrels.

A great number of new works are in progress along that stream. Good judges estimate the aggregate at fully one hundred. A village is rising at the mouth of the run.

THE RYND FARM.

That portion of the Rynd farm which has been subjected to the drill lies chiefly on the east side of the creek.

A machine-shop and a saw-mill are at work on the bottom. Of the whole number of wells sunk about one-fifth were active, when visited, and the proportion of new derricks to old is not very different—perhaps a little greater. The movement is generally made to higher ground.

Diamond Well.—Has been in operation two years. Pumps about twenty barrels daily. Depth, a little over five hundred feet, about the usual figure.

Rynd Farm Company's Well, No. Eight.—Has been going about two months. Yields, by pumping, twenty-five barrels per day.

Frost Company's Well.—Flows about five barrels per day.

Lathrop Well, No. Fourteen.—Testing. Show pronounced good. Product so far, estimated at five barrels per day. Well to be deepened. Owners evidently not satisfied with performance.

Lathrop Well, No. —.—Pumped two days, and said to have given about fifteen barrels per day.

Frost Company's Well, No. Sixteen.—Three years old. Pumped ten barrels per day until lately. Putting in new engine.

Frost Company's Well, No. —.—Used to give one hundred and fifty barrels per day by flowing; now gives ten by pumping.

WIDOW McCLINTOCK FARM.

There is considerable activity on this estate. Several wells which had lain idle, in addition to those mentioned underneath, are about to be deepened or start pumping again. Those in operation are the following:

Hammond Well.—Was a flowing spring. Produced five hundred barrels daily for a time. Is now idle. While tubes withdrawing, they fell to the bottom and could not be recovered. Proprietors now boring another alongside.

Chase Well.—Was pumping thirty-five barrels per day before the great freshet. Damaged by it and since idle. Manager clearing it out and preparing to resume work.

Bumstead Well.—Newly opened. Pumped one hundred barrels the first day—so stated. Seed-bags burst and not yet replaced. No reason to doubt the correctness of this statement.

McCue Well, No. Four.—Just reöpened. Yields half a barrel or so daily, but expected to do better when water is exhausted.

Georgiana Well.—Old work. Long idle, and just started again. Pumps about fifteen barrels per day.

King Well.—Belongs to Buffalo and Cherry Valley Oil Company. Been in operation about six months. Said to yield ten barrels per day. Outside statements and appearances incline me to put the amount at five or six.

Cincinnati Well.—Averaged fourteen barrels per day last week. Is expected to do rather better than *half* that quantity. Opened about first of January last, but not pumped regularly.

Westmoreland Well.—Just struck oil. Tools fell in and stuck fast. Good show of oil.

Painter Well.—About two years old. Averages about ten barrels per day.

Sterrett Well.—At one time yielded one hundred and fifty barrels per day. Now ranges from five to thirty, with an average of nearly twenty barrels. Day of visit, gave thirty.

Sterrett Well No. Three.—Struck oil about the first of

December last. Idle all winter. Now pumping about twenty-five barrels per day.

Sankey Well.—About eighteen months in operation. Flows about four barrels daily.

Ocean Well.—Yielded little before getting burned.

Yamhill Well.—In operation about six weeks. Giving six barrels per day, with steady yield.

ROUSEVILLE,

situated at the mouth of Cherry Run, which enters from the east, is one of the most active villages in the valley. It has, besides the post-office, hotels, etc., a bank, a church, and an oil refinery. Some travellers regard Rouseville as the metropolis of Petrolia, and if supreme disorder in laying out the place, mud, and high prices constitute the test, this village will take a high rank. It is named after Mr. Rouse, who died in one of the great conflagrations alluded to elsewhere. He was conscious long enough to bequeath large sums to charitable uses.

Of the works in Oil Creek valley, immediately in front of Rouseville, nearly seven-eighths were idle at the time of visiting. Along the bluff the ratio was not so unfavorable. About fifty new derricks have been erected on the heights on both sides of Cherry Run, at elevations from fifty to one hundred feet. These works are not all active. The wells on Cherry Run proper will be noticed under a separate head. Annexed are the statistics of the principal on Oil Creek:

Willoughby Well.—Was a flowing well three years ago. Now pumps about five barrels per day.

Shaft Well.—Belongs to Allen, Wright & Company.

Used to be a fifty-barrel well, but vein tapped by another, one hundred and twenty-five feet off. Now pumps from ten to forty barrels, according to condition of the tubes in other well. Up to last November, had produced twenty-four thousand barrels, worth probably one hundred thousand dollars. Uses its own gas in furnace.

Miller, Riddle & Company's Well.—Has been going since last January. Yielded two hundred barrels at first; now averages about sixty per day.

Faulkner Well.—Pumps about twenty barrels daily. Yield improving. Has been in operation since last December. Estimate above considered safe and rather low.

Manville Well.—In operation a little over three months. Pumps fully twenty barrels per day.

Gate Well.—Four months in operation. Yield averages ten barrels daily.

Rope Well.—Been going about three years, and still pumping about fifty barrels daily.

Haines & Anderson Well.—Was an eighty-barrel well until lately. Managers complain of bad luck, oil-vein having been struck by other wells, some as far as two hundred yards distant. Has done little for five months.

Rawling Well.—At one time, gave one hundred barrels per day; but suffers from same misfortune as the last-named. Pumped eighteen or twenty barrels the day before visit. Been in operation about six months.

Oil Exchange Company's Well, No. Two.—Yields very little.

Webster Well, (west side of creek,) *No. Eight.*—Put down last year. Gives from six to eight barrels daily.

Well on Lot No. Forty-seven.—Belongs to a Mr. Means. Produces six or seven barrels per day.

Chamberlain and Hibbard's Well.—Sunk last year. Average daily yield, ten barrels.

No. Forty-five Well.—Sunk last February. Yields about ten barrels per day.

THE BUCHANAN FARM

lies immediately below Rouseville, the works being nearly all on the east side of the creek. Like all the other bottoms low down the valley, this farm was severely scourged by the great freshet. At the time of my visit, four-fifths of the wells were idle; indeed, on the upper portion of it, *not one* engine was running. The greater part of the wells appeared to have been abandoned as bad jobs, not merely laid up by a temporary misfortune. It is possible, however, that, through an oversight of mine, two or three producing wells on this property have not been examined.

Clark and Banks's Wells.—One in operation about four years; the other, eighteen months. The two yield nearly one hundred barrels per day.

Willoughby Well.—Just started. Yields from one to two barrels daily.

Wadsworth Well.—Yields water only as yet.

Well on the west side creek, just started. Day before visit flowed some oil; but workmen cannot say how much.

THE JOHN AND HAM. M'CLINTOCK FARMS

come next in order, in passing down the valley. McClintock village is the "hub" of this "territory"—a feeble imitation of Oil City, squatting on the bottom, with a hotel, several boarding-houses, and two or three refineries. About one well in every five sunk was in operation when visited, and the new derricks bore a like proportion to old ones. Part of those have been erected over wells whose former head-gear had been washed away or destroyed by the flood. The principal degree of activity prevailing about the old works is nearly confined to the upper end of this property; but lower down there are others in progress.

McKinley Well.—Flows from fifty to sixty barrels per day. Stream never stops altogether, but increases and diminishes largely. Principal discharge takes place about once in six minutes.

Baltimore and Venango Oil Company's Well.—Situated on the east side. Oil struck last February. Pumped at the rate of sixty barrels per day, when managers pulled out sucker-rods and well began to *flow.* Now yields, according to information, two hundred and fifty barrels per day. Is probably good for two hundred barrels.

Hibbard Well.—Pumps about ten barrels per day.

Parker Well.—Close by last. Pumps one barrel daily.

Abbot Well.—In operation five months. Reported to be flowing one hundred and fifty barrels per day. Yielded two hundred and forty at one time. Is on the decline; and I estimate the actual yield at probably not far from one hundred and twenty.

McClintock Petroleum Company's Wells.—One had pumped thirty barrels per day for last ten days. Another yields about ten barrels.

A flowing well, on east side, (name not ascertained,) pours out a steady stream for some minutes, then subsides for about an equal space. Belongs to a New-York interest. Struck oil about a week before visit. Yielded sixty barrels per day at first, then about forty.

The success of these few really good wells in the lower end of the valley, at a time when its reputation had begun to wane, has caused property to appreciate once more; and adventurers are likely to turn their attention afresh to those bottoms.

THE CLAPP FARM,

lying still nearer to Oil City, is rather less productive and active than those above it. In some localities nearly every derrick is standing idle. About one-third of the works are in progress. Nearly all are on the west side, the bottom here being one-third of a mile across.

Hemlock Well.—On east side of creek. Before the freshet, pumped sixty barrels a day. Present yield not ascertained; estimate at thirty.

Cuba Well.—Been opened about eight months, and is pumping one hundred and fifty barrels per day, according to information. Estimate at one hundred and twenty.

Bliss Well, No. One.—Struck oil on first April. Flow estimated at fifty barrels per day.

Bliss Well, No. Two.—Just testing, with good prospects; but actual yield unknown. Quite wrathy with flow of gas and frequent spirts of oil. *Well No. Three* in progress.

Well re-tubed and just commenced again. Yield estimated at five barrels per day; gave one hundred and forty barrels at one time. Another well close by, pumping about two barrels daily. Names not ascertained.

On Cornplanter Run, entering from the west a short distance above Oil City, is a flowing or dripping well, whose yield (of lubricating oil) is from three to four barrels *per week.*

Between the mouth of this stream and Oil City are three other wells in operation, with an aggregate yield of about ten barrels per day.

On the last two miles of the valley, I counted about seventy derricks, a majority of which appeared to be abandoned. Signs of activity near fifteen or twenty. On and near Cornplanter Run were a dozen or so of works in progress. On or below the Clapp farm were nine refineries, of which only two were active, and several had been damaged by water. Found it very difficult to get facts thereabouts, as, in addition to the extensive solitudes existing, new hands had gone to work at several of the wells.

Between Oil City and Rouseville (three miles) the number of derricks standing was, last winter, ascertained to be four hundred and thirty-five, which would be equivalent to at least five hundred wells actually put down. In walking, multitudes of pits are seen without any upper works attached, these having been washed away, burned, or removed. As an approximate estimate, I give the number of wells put down on Oil Creek proper at two thousand, with five hundred more in progress. To these should be added one thousand more, either finished or going down, along the various streams discharging into that great oil artery. The new works will be found to number not far from one thousand in this basin only ; but the aggregate of abandoned ones is doubtless still larger. Assuming, then, that every derrick put up should be followed by its productive well, the whole number would be insufficient to replace those which have been sunk the past five years, and are likely to be left to themselves the present season. And, indeed, if we add to this basin those of the Alleghany, French Creek, Pithole, etc., I question whether the new will do more than replace all the old. But a comparatively small number of those put down in the past year, or to be put down in the present, will yield oil in paying quantities, much less gush forth as did the wonderful springs reached in 1861.

CHERRY RUN

is by far the most important tributary of Oil Creek. It is ten or twelve miles long; but at its mouth, scarcely more than six feet wide. At its lower extremity the valley is quite narrow, scarcely more than a stone-throw across, but higher up it widens, and the heights on each side become less abrupt. Above Plumer the run forks into two streams, along both of which large numbers of new works are rising. The wells already productive are confined to a space within two miles of its mouth; but when I visited the valley, several others approached completion near Plumer, and their managers felt confident of striking oil. It is scarcely a year since that secluded valley was invaded in good earnest by the petroleum interest, the Yankee well having been opened in the early part of 1864. This, followed by such celebrities as the Reed, the Mountain, the Auburn, etc., soon impelled a vast number of operators to the neighborhood, until the surface is literally forested with works, and the ground punctured every few rods. On one acre-lot, twenty-five derricks have already been erected. Beginning from the upper end of the valley, we find the following works in operation:

Rockwell Well.—Owned by Allen, Wright & Co.

Opened last June. Steady yield, pumping twelve barrels in twenty-four hours.

Dearborn Well. — Opened in middle of last July. Pumps fifteen to twenty barrels per day. Belongs to the Cherry Run Petroleum Company.

Carver Well.—Just tested. Pumps fifteen barrels per twenty-four hours. Yield said to be on the increase.

Emary Well.—Newly opened. Flows sixteen barrels, but will pump twenty per day. The latter tried occasionally. Yield said to be on the increase.

Main and Horn Bluff Well.—Flows one hundred and twenty barrels per day. Situated on the bluff, about fifty feet above the bottom. Six tanks, with capacity of about five thousand barrels.

Allen, Wright & Co.'s Well.—Has been flowing about a year. Yields an average of seventy barrels per day. Two others, belonging to same interest, pump respectively two and thirty-five barrels per day.

Follett Well, No. Three.—Been three or four months in operation. Yields about six barrels per day.

Anderson Well.—Owned by United States Oil Company. Reported to have been flowing three years (?) Now yielding eighteen barrels daily.

Ennis & Babcock Well.—Situated on face of bluff. Now giving ten barrels daily.

United States Well, No. Four.—Been in operation since last August. Yields twelve to fifteen barrels per day.

Potter and Jack Well.—Been in operation three years. Ranges from ten to fifteen barrels per day.

Utica Well.—In operation. Could get no facts.

Alhambra Well.—Belongs to Brown, Rockwell & Co. Pumped nearly one hundred barrels per day before burning down; now, about fifty barrels. Re-commenced about one week.

Auburn Refinery Well.—Owned by the Messrs. Orr, together with small refinery there. Well sunk eighteen months. Tools lost and just recovered. Raised three or four barrels lately. Prospect poor.

Alden, Brown & Perdu's Well.—Just started. Got six or eight barrels the day before visit.

Buffalo Well.—Has yielded in all about twenty-five barrels. Now dry.

Marietta Well.—In operation about four months. Yield irregular, on account of defects in machinery. Highest quantity, twenty-five barrels per day; average believed to be about fifteen for past week.

Well No. Nineteen.—Owned by Poole, Brothers. Never been fairly tested, but produced oil. Tools fast in bottom.

Union Petroleum Company's Wells.—Four already in operation. Aggregate yield, about one hundred and twenty-five barrels per day, or eight hundred and seventy-five per week. Three flow steadily and one irregularly. Been going various periods, from nine months to three weeks. Company will have five more completed in as many weeks; two others are also in progress. A New-York city interest.

Moody Wells.—*No. One* gives from twenty to twenty-five barrels per day and is gaining. In operation eight or nine months. *No. Two* is a new well, nearly equal to the other. *No. Three*, also new, only three weeks opened. Pumps nearly fifty barrels per day. Aggregate yield, probably eighty-five barrels.

Allen & Wright Oil Company's Well, No. Eighty-one.—Gives about twelve barrels per day, or eighty-four per week. *No. Eighty-four* gives forty barrels daily. Has been going about three months.

Brevoort's Company's Wells.—*No. One* pumps about fifty barrels a day. Four or five months in operation. *No. Two* flows, in irregular jets, about eight barrels. *No. Three* pumps about twenty barrels and is gaining. *No. Four*, about fifty-five barrels per day.

Mountain Well.—One of the greatest in all Petrolia. Is perched on the face of the bluff, fifty or sixty feet above the bottom. Very slight spasms perceptible in the discharges. Quantity yielded, given at three hundred and eighty barrels per day; but a person who "timed" it is positive that it took five days to fill a tank of eleven hundred barrels, averaging two hundred and twenty barrels. Its flow was considerably larger than that of Reed well, when visited. Very little water in either.

Reed Well.—Flowed one thousand barrels per day at first. Now reported at two hundred and eighty barrels. I estimate its actual amount at between one hundred and seventy-five and two hundred. Opened last July. Flow described as being uniform and steady, no change having been noticed during last three months. Is, however, undoubtedly on the decline. Well and two acres of land sold lately for six hundred and fifty thousand dollars.

Duff Tract or Crewell Well.—Owned by Mingo Oil Company of Philadelphia. Produces about fifty-five barrels per day, but rated at seventy. Recently opened. Oil comes spasmodically.

Bradley Wells.—One of them lately abandoned, after having pumped twenty to twenty-five barrels a day for some time. Fell off to nothing.

Rynd Farm Oil Company's Well.—Perched on bluff,

sixty or seventy feet above the stream. Depth, six hundred and fifty feet, or a correspondingly greater distance than those sunk in valley. Has been flowing about eighty-five barrels per day for four weeks.

Denny Well.—Belongs to Curtiss Oil Company. Pumps about sixteen barrels per day, on average.

Parker Well.—Testing. From appearances, they expect ten barrels per day.

Amazon Well.—Flows about twenty barrels daily. Open since last November.

Wood Well, No. One.—Just completed. Show considered rather poor. Other wells nearly finished on slope, said to have good show.

Rochester Well.—By pumping, gives about twenty barrels per day.

Baker Well.—Flowing from forty-five to eighty barrels daily. Safe to estimate the average at sixty barrels. Well reported at two hundred and twenty. Discharges spasmodically.

Watkins Well.—Flows and pumps, (to get the paraffine out of crevices.) Yield, about fifty barrels per day. Tubed during time of flood. Started week before visit.

Auburn Well.—Been going since last July. Yield, about sixty barrels per day, except when gas gave trouble. A flowing well.

No. Fifteen Well.—Started flowing last September. Ran up to one hundred and sixty barrels per day. Burst seed-bags and just started anew. This and Auburn foam, spirt, and flow spasmodically; while the Watkins pours out (by flowing) a steady stream for two or three minutes and then subsides. Estimate at forty barrels.

Ballot Well.—Flows about sixty barrels per day. Opened the present year.

Little Giant Well.—Yields about thirty-five barrels a day. Would give forty, if pumped steadily. Been four weeks in operation.

Wade No. Four Well.—Flows with a steady stream. Daily yield, about twenty-five barrels. Been going for two months.

Yankee Well.—One of the most eccentric productions of Petrolia. After twenty minutes' rest, discharging neither oil nor gas, the tube begins to emit gentle puffs, accompanied by little spirts of oil, both increasing for three minutes, until they became quite violent and frequent. Can be heard for a distance of two hundred yards. Noise and discharges then gradually subside, and at the end of five or six minutes stop altogether. Yield, about fifty barrels per day. This and several of the wells named above belong to the Cherry Valley Oil Company, who own a tract of land and have leased out several lots to other interests. The Yankee was the first grand success achieved in that part of the valley known as Smith's farm.

Apropos of that property, the correspondent quoted above writes: "Less than three years since, a man named Smith, poor in pocket and in resources, owned a farm [of fifty acres] about three miles up Cherry Run. Unable to get any thing out of his farm, he endeavored to sell it, and get down on the creek to try his fortune at oil-mining. His creditors were pressing and in a desperate mood he applied to J. W. Sherman, who had then recently struck oil, and was getting rich, offering to sell his farm for two hundred and fifty dollars. Sherman declined and advised him to keep the farm, in case something to his advantage might turn up. But Smith was determined to sell, and eventually found a purchaser at five hundred dollars. The new owner re-sold it at twenty-four hundred dollars, and just a year ago it was re-sold to the Cherry Run Oil Company for sixty-five hundred dollars. That company granted leasehold rights to bore on the land, reserving a royalty of half the oil. Only a small portion of the farm has yet been leased; but the royalty to the company is now three hundred barrels of oil, or about three thousand dollars daily. Thus two days' income about pays for the entire purchase of the property, on which the proprietors have not expended a dollar beyond the original cost."

I may add a few additional facts, obtained from the superintendent of that company. Of twenty-two wells sunk and tested on that farm, all but one gave petroleum in paying quantities, and even that exceptional case yielded a little. All the parties who have taken leases are represented as having succeeded, their net profits having ranged from ten thousand dollars to five hundred thousand dollars. There is no land for sale in that vicinity. Leases for quarter-acre lots are given for a royalty of one-half, and bonuses ranging from five hundred to three thousand dollars, the lessees engaging to actually sink one well.

Fee Well, No. One.—Near Yankee. Flows thirty barrels per day. *No. Two.*—Been running about seven months. Averaged fifteen or sixteen barrels per day till two weeks before visit, when got in an improved valve, and now yields twenty barrels.

Coleby Well.—In operation five months. Average

yield, by pumping, fifteen barrels per day. Seed-bag burst and had to stop for repairs.

Remmington or Gruningen Well, No. Eleven.—Started six months ago. Pumps about fifty barrels per day. Yield said to be on the increase.

Bradley Well, No. One.—Been going nearly eight months. Now averages ten barrels per day. Has been reported as a forty-barrel well.

Bradley Well, No. Two.—Yields about eight barrels daily. Opened since No. One.

Fauner Well.—In operation over a year. Average yield, twenty-five barrels per day, partly flowing.

Hicks Well.—New and testing. Good appearance of oil.

Mallory Well.—Six months in operation. Yield about twenty barrels per day. Flows occasionally. Outsiders give the actual yield at fifteen barrels. Was not yielding any, when visited.

In the bottom it has been customary to sink the wells about six hundred feet, and a proportionally greater depth, if higher up. As the stream has a rapid descent, this would bring the bottom of the Cherry Run basin about on a level with that of Oil Creek near Rouseville, where the wells average five hundred and fifty feet. The wondrous results obtained on the lower part of Cherry Run have led to most extensive operations all up to its headwaters. In six miles, the number of new derricks erected is at least three hundred. For half a mile above the mouth of that run, at the time of visiting, most of the works were idle—at least five out of every six—having been damaged by the flood or given out altogether. But the owners appeared to regard this comparative stagnation with indifference, provided the sixth well turned out handsomely. Further developments in and around Plumer will be looked for with great interest. Grouping together all the works actually tested in that valley, I estimate the profitable concerns at between thirty and forty per cent of the whole. The term "profitable" means that they return to their owner the actual outlay for sinking and management, interest on the capital, and the purchase-price of land at moderate (not speculative) values.

PITHOLE CREEK

is a tributary of the Alleghany, which it enters from the north-west, at a point about twelve miles above Oil City, after a course of as many miles. About half-way up, it parts into two branches, known as Big and Little Pithole, the latter rising on the elevated table-lands near the source of Cherry Run. The stream is small, and has a rapid descent. Except at its mouth, the works on which will be noticed under the head of the Alleghany River valley, the developments on Pithole have been quite recent, the first well having struck oil in January. The route usually taken to that section is through Rouseville and Plumer.

On Little Pithole, at the time of my visit, I was assured that a flowing well of over one hundred barrels per day had just been struck, a mile or two from the turnpike. It was out of my power to visit the work, and I give the statement (made by a mechanic) for what it is worth. Extensive preparations are making to sink wells on that stream. About forty derricks had already made their appearance, and large numbers of others were understood to have been contracted for. The Excelsior (New-

York) Company, it was stated, would sink fifty wells the present season. Riddle, Miller & Company had several on the way.

Along the head-waters of Big Pithole the same activity was manifest. One company had contracted to put down twenty-five wells on a single farm, this summer, such being one of the conditions of sale. The purchaser of land in the bottom was considered lucky, if he secured it at two thousand dollars per acre. At that place it would probably be necessary to sink seven hundred feet before reaching the oil-bearing rock. One well on the Dawson farm had reached a depth of five hundred and twenty feet, without any appearance. Along the principal roads, nearly one-half of the lands had been cultivated before the oil excitement.

Copeland Well.—Was struck on the eighth of April, at a depth of six hundred and forty-seven feet. Flowed sixty barrels a day at first; but defect in tubing discovered, and this had to be done over again. Working at this, when visited the well. Is certainly productive; but had not been going long enough to ascertain its actual yield. Belongs to United States Petroleum Company.

Lincoln Well.—On same farm and about six miles from mouth of stream. Depth, six hundred and forty-five feet. Yields twenty-five barrels per day, partly by flowing. Had reached the fourth sand-rock, (of course.) Another well in progress on same premises.

Frazier Well.—The making of Pithole, situated on the Holmden farm, nearly a mile below the last-named. Oil struck in the early part of January last, the flow being two hundred and twenty-five barrels per day at first. It is now rated at two hundred and thirty. Tanks have a capacity of eight thousand two hundred barrels, and were nearly full at time of visit. Learned that earlier in the season quantities had been taken off on sleighs. Just opening a wagon-road to the works. How lately the yield was tested I did not learn. The flow is continuous and uniform, with little water, being about one tenth the capacity of a two-inch tube. Belongs to the United States Petroleum Company of New-York.

Another flowing well has since been reported to have been struck on the creek, close by the Holmden farm. Rumor gives its product at one hundred and fifty and even two hundred barrels per day; but these figures are probably exaggerated.

The United States Petroleum Company are putting down six other wells on their tract at Pithole, besides three or four others elsewhere. According to appearances, one hundred new wells will be sunk in that valley the present season.

Near Wood's mill one was reported to have got a good show of oil, at the depth of seven hundred feet. That locality is about two miles below the Frazier well.

FRENCH CREEK.

This is probably the largest tributary received by the Alleghany, the creek being one hundred and fifty miles long, and one hundred yards broad at its mouth. Its general course is from north-west to south-east, corresponding with that of the Alleghany for fifty miles below Franklin. The stream is less sinuous than some others, while the valley is wider than most of the bottoms.

Boring for oil has extended to a point above the village of Utica, or ten miles from Franklin. The numerous transfers of land and current reports would indicate that before long openings will be made higher up-stream. On the tributaries of French Creek it is also probable that a considerable movement will take place before long.

In regard to the quantity, the quality, and the situation where oil is obtained, French Creek differs widely from Oil Creek and its tributaries. To notice them in order:

First. *The quality* is not the common illuminating oil, with a gravity of fifty or sixty degrees; but lubricating oil of nearly the best kind, its gravity being thirty to thirty-two; while the finest Mecca is twenty-eight degrees. Consequently it commands a much higher price than any other; while there is far less fluctuation in its market value. With the common variety in its crude state, ranging from three to twelve dollars per barrel, French Creek oil is ordinarily between twenty and twenty-five dollars. Yet the charges for transportation, and the internal revenue duties on both are alike. The French Creek kind has the additional advantage of being susceptible of use without any outlay or loss for refining, except when it is to be applied to delicate machinery.

Second. *In quantity* it also differs. For while several wells on Oil Creek and its tributaries have yielded from one hundred to three thousand barrels per day, the best on French Creek or (its tributary) Sugar Creek, has fallen below fifty barrels, while the average yield of profitable wells has not reached *five* barrels for a whole year. There is quite as large a proportion of idle and abandoned works along French Creek as anywhere else.

Third. *The situation* in which petroleum is obtained on French Creek is higher up geologically than elsewhere. Against wells which have to be sunk five hundred or six hundred feet along Oil Creek and Cherry Run, it is not customary to sink to a lower depth than three hundred, above Franklin. The second sand-rock there is reached at half the cost required to enter the third at five hundred feet. The only experiment that I heard of where oil had been obtained from the third sand-rock on French Creek, in paying quantities, resulted in bringing up the common article, not the kind got in the higher veins. This was at the depth of seven hundred feet, the increased distance arising from the general dip of all the rocks toward the south-west.

Fourth. The people of French Creek allege that they have one other advantage over their neighbors further north, in that *the proportion* of wells actually paying expenses is larger there than elsewhere. They claim that a majority of all those put down have repaid both the original outlay and the cost of management. Some even put the ratio of profitable works as high as six or seven out of every ten. My belief is, after making due inquiry, that both these figures are a good deal too high, and that three or four out of ten would be much nearer the mark. This is certainly a liberal enough estimate. It is true that the receipts for one barrel per day will cover all running expenses; while a second barrel, if contributed for only twelve months, will return the principal invested and a liberal interest. Still, it must be borne in mind that not one well in a hundred keeps up its accustomed yield day after day for a whole year, interruptions of various kinds happening frequently at the best seasons.

But with fewer first-class prizes to offer than other localities, though there might be as much diminished risk, it has come to pass that French Creek has drawn to itself comparatively little attention; and the number of new as well as old works may be reckoned in tens, instead of hundreds. The price of lands varies from one thousand dollars per acre, near Franklin, to one hundred dollars, near the sources of Sugar Creek. The plough and the hoe are not yet relics of a past age along these valleys.

In going down the railroad from Meadville to Franklin, the visitor passes seven idle or unfinished wells before reaching the first in operation, namely:

Henrietta Well.—Present yield, three barrels per day. Been nearly three years in operation. Flowed for a time. Belongs to a Philadelphia interest.

Of the next nine wells six are in progress. One has been entirely abandoned, the machinery being removed. One was stopped by the flood, after going two years. It was yielding two barrels per day at the time. The last gave three barrels daily; but complaint was made of the engine as being too weak. The tank had also been washed away. All standing idle.

The next half-dozen works consist of two abandoned wells and four in progress. Following these, we have four belonging to the French Creek Lubricating-Oil Company, described as follows: *No. One* gave two barrels per day for five months, but was stopped by the freshet. *No. Two* gave a very small quantity. *No. Three* is an old well, having been in operation four years. Was yielding five barrels per day before freshet. *No. Four* was working its way downward in search of the third sand-rock, but at the depth of six hundred and fifty feet had not found it.

One well (proprietors unknown) was in operation, and said to be giving three barrels per day.

The Pierson Oil Company of Philadelphia own five

wells. *No. One* yielded in all about twenty-five barrels, several years since. Now idle. *No. Two* gives one barrel per day, against four barrels, its former product. None of the others is doing as well as *No. Two.*

The McCormick and McKissock Wells are on the south side of the creek. *No. One,* very old concern, and reported to pump about one barrel per day. *No. Two* is new and pumps about same quantity in three hours, resting other twenty-one. *No. Three* in progress.

Neidler Well.—Flowed about twelve barrels a day for eight weeks, and subsided.

Another well (name unknown) pumps four barrels a day. Used to give six. Been in operation nearly two years.

Immediately above Franklin are nearly a dozen works in progress, and half as many idle or abandoned. Of these, some are reported to have yielded three or four barrels a day, and probably repaid their cost. Between Utica and Franklin there was no bridge, and my examination was confined to the north side of the creek. On the opposite bank, however, it was too plain that four-fifths of the works had been doing nothing for some time. That portion of this chapter relating to details along French Creek will be found less full and satisfactory than I could desire.

The Dale Oil Works is the name of a neat refinery about one mile above Franklin, on the north side of French Creek. Went into operation in the autumn of 1864. Number of stills, two; capacity of works, three hundred and fifty barrels per week.

SUGAR CREEK

is a tributary of French Creek, into which it enters from the north. It is too large a stream to be easily forded, and the bridge near its mouth had been swept away by the flood, so that crossing on foot had to be performed by means of trees thrown down in wild-cat fashion. This stream also parts in twain, the forks taking their rise in the plateau beyond Corry. Between the mouth of Sugar Creek and Coopertown—five and a half miles—about twenty-five derricks are already up or rising, but at these only one well has been thoroughly tested. It is two miles up-stream, and has yielded as much as thirty-five barrels a day for a month in succession; but through derangements to the machinery, it is supposed the last month's average did not exceed fifteen barrels. The well was struck about January last. The article is best quality of lubricating oil, of thirty-two gravity. The effect of this fine strike has been to enhance largely the price of real estate along the valley. One farmer, close by, refused an offer of one hundred and fifty thousand dollars for eighty acres, about half-bottom lands. A report, since my visit, is spread that a new well has been struck close by, and this is impelling a wave of speculation into that secluded and pleasant neighborhood. It is reported that lands are changing ownership as far as ten miles up. Probably one hundred new wells will be opened on that creek before the close of 1865; a number of old ones will also be burnished up anew or sunk deeper.

FRANKLIN.

The wells in Franklin have all the characteristics of those on French Creek. Only a few of them are now in operation, but the greater number would appear to have paid for themselves, and a few handsomely.

Lamberton's Well has been in operation nearly four

years. Yields three and a half to four barrels per day. Owned by Chicago and Alleghany Oil Company.

Ingall's Well, (*Caledonian.*)—Idle at present. Yielded three to three and a half barrels a day, when stopped.

Star Well.—Yielded about same quantity. Operated about one year after opening in 1860. Stopped. Leased to another party, but with no good result.

Cairns' Well.—Nothing done for two years, but testing with a view to resuming operations. Sunk in 1859.

Cooper Well.—Old concern also. Owned by a company in Franklin. At one time pumped three to four barrels per day.

Dale Well.—Another of the ancients. Gave two to three barrels per day for a while. Upper works carried away by freshet.

Mammoth Well.—At first flowed eleven barrels per day, but fell down to four. Been idle for some time before flood, when its works were carried off.

Evans Well.—Belongs to Moseley & Company, new owners. At first, yielded forty barrels a day, which turned the head of *one* fair damsel. Yield has since declined to five barrels. In 1861, operations were suspended, and the well remained idle until last winter.

Broomstick Wells.—Belong to Forest Shade Oil Company of Philadelphia. Two in number. Are about four years old, and yield four to five barrels a day each.

Bonnell Well.—Leased by Moseley & Company. Old concern. Deepened, abandoned, recommenced, and now deepening a second time. Indications pronounced good.

Besides these are about half a dozen old concerns thrown up, whose history and production could not be ascertained. Most of the wells put down during the early stages of the oil-fever are in the upper part of the borough. At the lower extremity are several new ones, of which particulars may be found elsewhere.

THE ALLEGHANY, BELOW FRANKLIN, EAST SIDE.

My examination of the Lower Alleghany valley extended to East-Sandy Creek, and some distance up and beyond it. The following summary of wells on the east side of the river will be interesting: Abandoned, eight; idle, temporarily, twenty-four; active, thirteen; in progress, as many by the river, and (probably) fifty more on Two-Mile Run and East-Sandy. Beyond the mouth of East-Sandy, I could not learn that there was more than one well in operation which yielded oil in paying quantity. Both old and new derricks, however, may be seen for some distance further. On both sides, the river-banks are steep, lofty, and close to the Alleghany.

The quality obtained is first-class illuminating-oil, the lubricating kind not "cropping out" below Franklin; though it does for a short distance above, along the great river. Crossing on the fine suspension-bridge, we first find—

John H. Lee's Well.—Opened in March, about four weeks before visit. Flowed at first over two hundred barrels per day, but this gradually subsided, and owner took to pumping. Present yield, twenty barrels daily, working twelve hours. Was idle at the time for lack of facilities to carry off the oil. Tank filled. Depth of well, four hundred and forty-eight feet.

On some wells close by the Lee, preparations for an active campaign are going forward, with fair prospects in some. Along the bottoms, for some distance beyond, are numerous wrecks and other "indications" of damage by the great freshet. One well has been filled up with mud and is useless. One has tools fast in bottom, after pumping twenty-five barrels per day for a time.

Two-Mile Run.—Judging from appearances at its lower end and the reports of others, I estimate the total number of wells in progress on this stream at twenty or twenty-five, of which only one has been tested. Said to yield three or four barrels per day, and on the increase.

MEM.—This brook must not be confounded with a Two-Mile Run which enters the Alleghany the same distance *above* Franklin, and from the opposite side.

Keystone Well.—On Cochrane farm. Put down about two years ago. At first yielded about seventy barrels per day, but gradually fell off to twelve, when the floods came and the winds blew and beat upon that well and filled it up.

Williams and Moyer Well.—In operation. Has pumped six to eight barrels per day for a short time.

Dale and Morrow Wells.—Two in number. Pump about thirteen barrels per day in all. One has been going two years; the other eight or nine months. Proprietors sinking others along the bluff.

Painter Well.—In operation five weeks. Yields two barrels per day—never more than three. Has passed as a fifteen-barrel well.

Pennsylvania Oil Company's Wells.—Four in number. Two have yielded three or four barrels each; the others not yet tested.

Alleghany and East-Sandy Oil Company's Wells.—Two old concerns, sinking deeper.

Island Well.—Said to be giving from twelve to fifteen

barrels per day. I credit the estimate as not beyond the mark.

Robertson Mining and Oil Company's Wells.—No. One yields eight to nine barrels per day. Four others give average of one barrel per day each. *No. Two* gives two barrels. Aggregate yield, when going, about fifteen barrels. For lack of fuel, all were standing idle. Plenty of cord-wood close by. Some of these wells belong to the early period.

Greenhill Oil Company's Wells.—Three in number. Two of them new. Old one yields one to one and a half barrels per day. Others not thoroughly tested, but one promises fairly. All pumped "by head"—two or three hours per day. Depth, about four hundred and forty feet, but to be sunk deeper. General depth on Lower Alleghany, from five hundred to five hundred and fifty feet.

Sunbury (or Sonsberry) Oil Company.—Five wells sunk, and one in progress. *No. One*, sold lately as an eight-barrel well, yields one to one and a half. *Prospect well* gives three barrels. *Smoky City*, three barrels; *Heidelberg*, eight to ten, but now idle; *Kendrick well*, probably ten. The last at one time yielded one hundred barrels daily. All have been two or three years in operation. Not pumped steadily.

MEM.—One of those wells, formerly known as the Blakeley, was lately sold for twenty thousand dollars, as an eight-barrel well, the average yield being about *one!*

Crozier Oil Company's Well.—Idle, tools having stuck fast. Pumped oil, but not in paying quantity. Depth, six hundred and fifteen feet. Talk of going still further down.

Hope Well.—Belongs to Superior Oil Company of Pittsburgh. Sunk in 1861, and deepened last year to nearly eight hundred feet. Yield estimated at four barrels per day, before the flood. Filled up with mud. Well to be re-reamed and widened.

Porter Farm Oil Company's Wells. — *Morris* well abandoned. Engine said to have blown up after testing. Probably remains idle because the spring did *not* blow up. Yielded oil, but quantity unknown. *No. Four* testing, with good show, at depth of three hundred and fifteen feet. Another stopped up and wrecked by freshet. Another sunk over six hundred feet. Obtained oil, but in insufficient quantity, and tunnel to be extended toward "China."

EAST-SANDY CREEK.—Enters Alleghany River from east side, six miles below Franklin. Current quite violent, and stream one hundred feet wide at its mouth. Bridge swept away, and "Charon" charges ten cents for ferrying across. Two wells tested on the lower three miles of its course, namely:

Adamantine Oil Company's Well, No. Two.—Sunk four hundred and fifteen feet, when tools got fast.

Soft Maple Well.—Owned by same company. Depth, four hundred and twenty-seven feet. Has been two months in operation. At first, pumped one hundred and fifty barrels per day; now gives eighteen barrels, and keeping steady. Tubing said to be somewhat out of order.

Keystone Well.—About three miles up creek. Yields one barrel per day, but quantity reported at four or five. Facts learned from a source deemed trustworthy.

One well on East-Sandy has been sunk eight hundred and fifty feet, without finding oil, except a small show at the depth of three hundred and forty feet. From fifteen to twenty are in progress on the creek.

8

One well, lower down the Alleghany, on same side, sunk six hundred and twenty-eight feet, yields eight barrels per day. Formerly owned by Iron City Oil Company. Present owner's name unknown to my informant.

LOWER ALLEGHANY—WEST SIDE.

Starting from the mouth of West-Sandy Creek, we have, on the Miller farm, two wells pumping about four barrels each. Two others going down. On the Foster farm no wells producing oil. Three in progress.

Excelsior (N. Y.) Oil Company's Wells.—Situated on the D. Smith farm. Three fully tested and one sinking. *No. One* pumps six to eight barrels per day. Another gives three to five barrels, with average of four. A third was yielding ten barrels before, in making repairs, the tools stuck fast. Is not now paying expenses. A fourth in progress.

Hubbs Well.—Now idle. Sold as a sixteen-barrel well last fall. Yielded about twelve barrels, when stopped.

MEM.—Many of the wells below Franklin were sunk in 1860 and abandoned afterward, when oil sunk in price to an unremunerative point.

Excelsior (Philadelphia) Oil Company's Wells.—Two in number. One yields two or three barrels daily. The other is in progress.

Overton Oil Company's Wells.—Two, both leased from Excelsior Company. One yields two and a half barrels daily ; other re-tubing. Is known as the *Childs well.* Used to give five or six barrels. Depth, four hundred and sixty-seven feet. Been in operation four years. A third well in progress.

Steppy Farm Well.—Put down in 1859. Yields one

and a half barrels per day. Flows a little at times. Has not been regularly operated lately.

MEM.—Two wells in progress; one, belonging to a Pittsburgh interest, with good show at five hundred and eighty feet. Three others sunk and now idle. One gave about a barrel daily. Others *had* yielded oil, but were thrown up. One was sunk only two hundred feet. Belong to Pope Farm Oil Company.

Organic Oil Company's Wells.—Two in number, each pumped two years, "by head," about one barrel per day. Both idle. Depth, five hundred and six hundred feet.

Thompson Well.—Owned by Pope Farm Oil Company. Depth, five hundred and forty feet. Opened four years since, but never tubed till last fall. Pumped for a time ten barrels, but has fallen off to six barrels in twenty-four hours.

Ravine Well.—Also on Pope farm. Sunk in 1860. Depth, four hundred and fifty feet. Before stopping, yielded about five barrels per day. Has not been running for some time.

Blakeley Oil Company's Wells.—Two in number. Both in progress. One is sunk four hundred and fifty feet.

Dr. Pancost's Well.—Now idle. Produced six barrels per day some time since.

Harvey Evans's Well.—On Hoover farm. Gave three barrels per day before stopping, last winter.

Passed five works. One wrecked by flood. One tested, but gave nothing. Others untested.

James Grahame's Well.—Damaged by flood. Flowed three barrels per day at one time; now, one and a half. Above this are four wells in progress.

Dixie Well.—Pumps six barrels per day. Used to

give ten. Four years old, and seven hundred feet deep. Most of the oil got at four hundred and thirty feet, in the second rock.

Apollo Well.—Running four years, and four hundred and forty feet deep. Stopped from 1862 till last year. At first gave twenty-five barrels per day, but declined to zero. Re-tubed, and lately yielding ten barrels. When visited, was idle for lack of fuel. Four wells in progress immediately above.

Buyer Oil Company's Well.—Pumps two barrels per day by head. Depth, five hundred and fifty feet. Sunk in 1864. Never yielded much. Been rather improved by re-tubing.

Hoover and Marshal Well.—Depth, four hundred and fifty feet. Four or five years old. At first gave twenty-five barrels per day. Now pumps from eight to fifteen, with average about twelve.

Catfish Well.—Is pumped by head, (a few hours daily.) Actual yield, four and a half barrels per day. In operation nearly two years. Depth, four hundred and forty feet. At first, gave eighty barrels. One other well in progress on lot.

Rhinehardt Well. — Sunk nearly three years since. Gave two hundred barrels per day at first. Lately pumped twenty-five barrels. Now re-tubing. Has been most productive on that side of river below Franklin.

Hoover and Marshal Well, No. —.—Pumps about twenty barrels per day. Another, belonging to same company, idle, having lost tools in bottom. Did yield oil, but statistics unknown.

MEM.—Every well tested on the Hoover farm is said to

have yielded some petroleum. The present owners charge a royalty of one-half the oil from lessees.

Hemlock Well.—On Lee farm. Gave twenty barrels per day at first; pumped two or three barrels lately. Out of order. Mem.—On the lower part of this farm are three idle wells, one apparently thrown up.

Old Lee Well, and another adjoining it, belong to a Boston company. Both productive, but idle, undergoing repairs. Former rated at twenty-five, latter at seven barrels per day, when in operation.

Suffolk and Venango County Company's Wells.—Two in number. Can be made to yield fifty barrels per day, when in proper order. Estimate somewhat vague. Learn that actual yield of one has been twelve to fourteen barrels; of the other, five to seven barrels daily.

Honeycomb Company's Well. — Also on Lee farm. Flowing thirty-five to sixty barrels a day. Average believed to be nearly forty. Six months in operation. Keeps steadily up in yield. Depth, four hundred and fifty-five feet.

Wallace's Well.—Upper works swept away. Was yielding seven barrels daily before the flood.

Sprogel & Co.'s Well.—Depth, four hundred and fifty feet. Average yield for year, five barrels per day. Is eighteen months old, and now known as Alleghany Company No. Two well.

Alleghany Company's Well, No. One.—A two-barrel concern. In operation over two years.

Westminster Well.—A one-barrel spring. Six months in operation. Two others were sinking. Both damaged by flood and now idle.

Eureka Well.—On Morrison farm, a short distance below Franklin. Old work re-reamed. Preparing to start anew.

Passed three wells idle and seemingly abandoned. Two of them yielded about three barrels a day each; the other, nothing. Three other wells in progress.

Thomas Well.—Bought as a thirty-barrel concern. Report of its having been fed from a tank on the sly pronounced false. Pumps about two barrels per day. An old work, and never of much account.

Passed six more wells sinking, and entered Franklin. The journey altogether the roughest and toughest I had then undertaken. The river was high, and no road existed save a goat's path between the precipice and the flood. So it is likely to continue.

ALLEGHANY RIVER—FRANKLIN TO OIL CITY.

The distance between those points has been increased a full mile by the destruction of the lower bridge across French Creek. On the lower end of this creek, passed seven wells in progress, two abandoned, and one wrecked.

Stock Well.—On the Alleghany. Pumps seven barrels per day best quality lubricating oil. Gravity said to be twenty-eight. Depth, two hundred and seventy-five feet.

Martin Well.—Pumps five barrels per day of same quality and at same depth. Both wells over four years in operation.

Passed two wells, said to have been abandoned when oil was low, though yielding tolerably. Leases expired, and property reverted to owner—a matter of quite frequent occurrence under the old dispensation. Counted eleven more works, with or without derricks, all seemingly given up.

Enterprise Oil Company's Wells.—A Pittsburgh interest. Before the flood, *No. One* was yielding five barrels per day. *No. Two*, about four barrels. Both seriously damaged, but preparing to resume. Three or four years in operation.

Plumer Refinery.—Belongs to a New-York interest. Expected to go into operation shortly. One well on premises idle. Some others to be put down.

United Farm Company.—Own large tract of bottom-land below the mouth of Two-Mile Run, *above* Franklin and on west side of river. Have leased lots to several interests. Twelve new derricks already erected this season, and about thirty new wells expected to be sunk. Flat greatly damaged by freshet, and all the old works suspended. It is said that none of these were fairly tested; hence the yield was unsatisfactory.

On Two-Mile Run.—One old well bored out again with fine appearance, as reported. On that little stream, one hundred wells supposed to be sunk or sinking. Such estimates are usually exaggerated one-half, by mistake.

Powell Oil and Coal Company.—Own six wells in progress on Shirk farm. Most of these bored three or four years ago, but now deepening. In some, a good show reported.

Frankford Oil Company's Wells.—Three in number. Two idle; one pumps average of sixty barrels per day. Opened on first of March. Is situated about midway between Franklin and Oil City; and if the figures only keep on thus, will make princely fortunes. Another well, belonging to same company, *drips* one barrel per day. Upper works of a third were swept away. Discharge from the *big* well is equal to a stream of two hundred barrels per day; but when visited, nine tenths of it seemed to be *water.* Workmen say, at times it is nearly pure petroleum. No regularity of flow. Depth, five hundred and thirty feet.

Pembroke Oil Company's Wells, No. One.—Yields about five barrels per day. Been keeping up to that figure two years. Depth, three hundred and eighty feet. Company have four more under way—one tested, but not producing oil.

Within the space of a mile above these occur seven wells, four of them wrecked, one standing idle, one in progress, one just tested and said to be producing.

At Reno Station, on the railroad, are three refineries—E. W. Shippen's, Kincaid, Lockwood & Co.'s, and Ward & Lockwood's. The largest has four stills and a capacity of two hundred and seventy barrels per week. One has abandoned the refining of illuminating oil as unprofitable, and taken to the lubricating kind only.

Howe & Eddy Oil Company's Wells.—Three owned and three leased. Only one going, with average yield of two barrels per day. Another pumped three barrels before freshet.

Kifer Elliott Well.—Just opened. Appearance and prospect considered good. Pumped at rate of twenty barrels per day for short time. Depth, four hundred and ten feet.

Between that point and the outskirts of Oil City—about a mile—counted twenty-two wells, only four of which were going or testing. Eight were idle and ten sinking. None of those yielding gave over three barrels per day. On the opposite side of the river, between Oil City and Franklin observed about forty derricks, of which nearly one-half had been recently erected. Wells unfinished, and at many no progress. Only three or four were actually producing oil. Statistics not learned, but never heard that section of Petrolia referred to as having aught worth visiting. No regular means of crossing.

OIL CITY.

Chris Kringle Well.—Going since last Christmas, when yield was thirty barrels per day. Now pumps twenty barrels, according to official estimate. When visited, was discharging gas copiously, but no oil. Some good wells are in the habit of indulging in such pranks before strangers. Depth, five hundred and fifty-five feet.

Oil City Petroleum Refinery.—Capacity of works noticed elsewhere. Company have two wells in progress. Another refinery stands close by.

Harper Farm Well.—Pumped eight or nine barrels per day for past week. Old concern, deepened to five hundred feet.

Sweeney Well.—Official report of yield, twenty barrels per day. Others say from eight to ten. I incline to the latter estimate. An old well, deepened to five hundred and forty feet. Been three months at work since reöpening.

Shirk Well.—Tested only a few days. Flowed at first thirty-six barrels per day. Since the flood, average yield not above ten or twelve, got by pumping.

Old Glyde Well.—Belongs to Harper Farm Company. Been going two years, and now yields three barrels per day.

Linden Well.—Noted as a flowing well. Until recently, yielded ten barrels per day. Re-tubing at time of visit.

Glyde Well, No. 1.—New concern. Pumped five barrels first half day. Appearance good.

In that part of Oil City below the creek, are about thirty wells in progress, some in the ravine between the heights, but the greater number between the street and the river. Two or three old works are also being deepened and burnished up anew.

THE ALLEGHANY VALLEY—ABOVE OIL CITY.

My explorations of that part of the Alleghany valley between Oil City and Irvine (fifty miles) were conducted partly on foot, partly by skiff or raft, and the residue by steamer. Among the several modes of conveyance in Petrolia, I feel it my duty to characterize the steamer "Advance," as furnishing the most costly and uncomfortable. In charging as fare three dollars for twenty miles; in dirt, discomfort, package, discourtesy, noise, to say nothing of indigestible viands, it may suffice to say that all was of a piece. Even the time lost in waiting till the wind lowered, (!) and afterward at landings, was such, that the whole distance could have been easily walked in the same time.

On the line of Spring Creek, and on that of the Brokenstraw, into which the former discharges, there is little doing in searching for petroleum, the number of derricks visible from the railroad being barely half a dozen, between Corry and Irvine. The proprietor of the flats at Irvine is said to have set his face fixedly against any desecration of them by boring for oil; hence the derrick does not obtrude its gaunt visage thereabouts. It is only on approaching the village of TIDEOUTE, fourteen miles down the Alleghany, that the stranger finds himself once more within Petrolia. The principal works there are the—

Economy Wells.—These belong to an association of Christians known as Rappists or Communists, established on the common-property principle, which distinguished the apostolic church in Jerusalem. More than half a century ago, a society of Germans migrated to the United States, and founded the village of Economy on the Ohio River, a short distance below Pittsburgh. Subsequently they removed to the Wabash valley; but returned to their old settlement on the Ohio, where they still remain, owning property, the value of which has been estimated at the round sum of ten millions of dollars! The most objectionable feature in this association is the enforcement of the practice of celibacy, as practised by the Shakers. Still, its ranks are recruited from the outside world, until its membership counts by hundreds.

The way this society came to be concerned in petroleum operations was this: They had lent money to a farmer living near Tideoute, who failed to fulfil his obligation of payment; and his property consequently passed into the creditors' possession. About that time the developments on Oil Creek began to attract attention; and as springs of petroleum had been found at the surface on their property, the society decided to put down a well for themselves. In 1861 the first was sunk, and a fine quality of illuminating oil struck, at the depth of one hundred and thirty-five feet. This received the appellation of the "A well," which is still pumping five barrels per day, (twenty-four hours,) besides supplying gas sufficient to keep two engines running. "B well" gave feeble indications at first; but gradually improved, and now pumps from thirty to thirty-five barrels per day. "C" produces about fifteen barrels in the same time, and "D," from fifteen to twenty barrels. In none of the last three has there been lately any perceptible increase or decrease. None of them has been sunk over one hundred and forty-five feet, which makes such a copious and continuous yield the more remarkable.

At one time, the society leased the property, under conditions deemed favorable, to the former proprietor, who, according to outside reports, held his grip upon it till the society came down with an immense sum to recover possession, while the works went to wreck. New derricks, engines, etc., have since been erected, together with an excellent boarding-house, whose arrangements are unquestionably the best in all Petrolia. The restrictions placed upon employés, however, in regard to speech and manners, have made the proprietors somewhat unpopular as employers; and many prefer to seek employment where they "can blow off steam," without meeting a printed rule or other restriction upon their language or demeanor.

The lands belonging to this association on the Alleghany comprise eight thousand acres, extending back a long distance from the river, and having an extensive water-front. Much of this tract is covered with a heavy growth of timber, almost as valuable as petroleum these days. A saw-mill and a flouring-mill, driven by steam, have been erected on the uplands. A railroad is also in contemplation, to connect their various works along the river with a landing. Arrangements are in progress to put down ten or twelve new wells the present season, on spots where good indications are said to exist.

For the information of the curious, it may be proper to

add that no part of this property is either for sale or to let. Land speculators and persons engaged in getting up oil companies may, therefore, as well give it a wide berth in their searches after "first-class territory."

Those who delight to trace every unusual event or phenomenon to "special providences" may here find profitable matter for reflection. Why should the only good tract on that side of the river, along that part of its course, and one which has produced the most enduring springs in the country, have passed into the hands of an association having "all things common"?

Below the Economy wells are half a dozen idle works, most of which appear to have been thrown up altogether. One of them, known as the Rawlson well, belongs to a Philadelphia company, which had it sunk to the depth of one thousand feet, without finding any oil worth mention. Only one of the whole number is estimated to have paid expenses. The Waters & Jackson well ran seven or eight months, in 1861, and gave as many barrels per day. It is now idle. Three of them flowed for a few weeks at first.

Hockenburgh Well.—Named after a clergyman in that neighborhood, who has written an essay on the all-absorbing subject. Pumped about ten barrels per day for nine months, in 1861; then shut down. Now owned by the Tideoute Bayou Petroleum Company, of New-York, and yielding five or six barrels per day.

G. I. Stowe Well.—Sunk in 1860. Yield, four or five barrels per day till last September, since which has been idle. Depth, one hundred and forty feet.

Moore, Blanche & Company's Well.—Pumped five or six barrels per day until lately. Now undergoing repairs.

Towner & Thompson Well.—Pumps four to five barrels per day. Some alterations going on. Tried the blower, but unsuccessfully.

About eight more wells, all idle, and as many others in progress, are on the east side of the Alleghany at Tideoute. Above the village are also twelve or fifteen abandoned works, not having been going for some years.

On the opposite side, where the river impinges upon the bluff, counted sixteen new derricks, several of them perched at points from one hundred to two hundred feet up the precipice. Immediately beneath them are the following:

Moser Well.—Just resuscitated after a winter's rest. Yields six barrels per day. Sunk four or five years ago.

Rawlson Well.—Pumps eight barrels per day. Over four years old. Depth, one hundred and fifty feet.

Gilson Well.—Owned by Tompkins County and Tideoute Oil Company. Pumps thirty barrels per day. Opened last Wednesday. Has increased in productiveness lately. Depth, one hundred and forty-six feet.

Shaw Well.—Pumped, during winter, ten barrels per day, to best of informant's knowledge. Now stopped for repairs. This and four or five others to recommence shortly. About a dozen appear to be abandoned altogether.

The Tideoute and Warren Oil Company own most of the lands in that part of the village or its suburbs. They grant leases of lots for fifty per cent of the oil. That locality is fast following the example of Oil City, as respects filth and disorder, the latter place being the unit of measurement for the whole region.

Below Tideoute passed twelve or fifteen derricks, all old and inactive, until arriving at

HICKORY CREEKS.

These are three in number, all entering the river within a distance of two miles. The highest up of these is East-Hickory, at the mouth of which is a good landing, with half a dozen wells in different stages of construction. One of them is reported to have reached the depth of four hundred and thirty feet, where a good show was lately had in the sand-pump.

Little Hickory enters the Alleghany from the same side. Five or six wells going down on the flat at its mouth; but no definite result had been obtained at the time of visiting.

West-Hickory enters nearly opposite to the last-named, and from the west side. All have dug their channels deeply into the bluffs; but the only large bottom is near the last-named. At the distance of barely half a mile from the river is

Hickory Flat Well.—Oil recently struck at the depth of one hundred and ten feet. Came up so violently as to drive away the workmen. Has been flowing at the rate of one hundred and seventy barrels per day, or one hundred and ninety-six barrels in twenty-five hours, as ascertained by actual measurement. Oil said to be finest quality lubricating with less than thirty degrees gravity. The flow is steady, and the stream refreshing to behold, accompanied by gas in moderate qnantity.

The success of this well has given a powerful impetus to similar enterprises in that part of Petrolia. Already much the larger portion of the land, for miles above and below, has changed ownership. One farm of four hundred acres was lately sold for one hundred and twenty-nine thousand dollars, which may be taken as a fair estimate of price, when sold in large quantities and near the river. It is said that a speculator recently cleared sixty-six thousand dollars by a single purchase. Natural oil-springs, Mr. G. A. Siggins assures me, had long before been observed thereabouts, and occasional rude efforts been made by the residents to reach them—in one case, by means of a drilling apparatus composed of a bed-cord and a harrow-pin! The Indians had a reservation close by, which they inhabited until a few years ago. Contiguous to this well are aboriginal remains of various kinds. A large tree of a wood resembling oak (not petrified) was found at the depth of forty-nine feet below the surface, after drilling through a bed of solid rock.

About half a dozen new wells, one of them productive, were under way along West-Hickory, at the time of visiting. Preparations were going forward for extended operations during the summer, and that valley is now a favorite of those who are in search of new "territory."

At Dawson's crossing, lower down the Alleghany, two wells are in progress, one having reached the depth of two hundred feet.

At TIONESTA, an incipient village at the mouth of Tionesta Creek, and some twenty miles above Oil City, seventeen derricks have been erected at or near the mouth of that stream. One well has been sunk to the depth of seven hundred feet, without finding oil. Two or three works on the opposite side, but with no better success as yet.

Between Tionesta and Walnut Bend, fourteen miles, my inquiries were not so full as I could wish, having been made only at those points where the steamer stopped on her way down. The residue of the distance to Oil City I had previously traversed on foot.

PRESIDENT is a new and growing village on the east side of the river, with a capacious hotel and store as the nucleus. About two dozen derricks stand on the gentle slope, with abundant room for more. Two thirds of these works are unfinished. One well only was pumping oil, at time of visit. Said to be yielding freely, but figures not obtainable. A passenger dubbed the smaller collection of houses, etc., on the opposite shore, *Vice*-President. The next point of interest is the mouth of PITHOLE CREEK, twelve miles above Oil City.

Ewing Well.—Pumps fifteen barrels per day. Was lately bought as a twenty-barrel well by the Niagara Falls and Cherry Run Oil Company. Been four years in operation, and yields steadily. Oil discharges in occasional spirts, not a constant stream. Depth, two hundred and ninety feet. Gravity, forty-four.

About thirty wells have been sunk near the mouth of Pithole Creek, chiefly along the Alleghany. Of these, only one is now in operation. Owners are generally preparing to deepen them, or throw them up altogether. Taking both sides of the river and the works in progress, the number of derricks is not short of fifty. Much activity exists at the landing, and it is fast taking position as a village.

At the next bend below counted sixteen derricks, all idle and mostly new. Saturday evening, and the men may have quit work.

At Walnut bend are nearly as many, the greater portion finished works, and some quite ancient, as men reckon antiquity in Petrolia. One concern at work, and reported to produce finely—in fact, has made the reputation of that place. A good deal of activity on that semi-circular headland at ordinary times. Is on east side of the Alleghany, and nearly eight miles above Oil City. On the west shore are the following wells:

Black Diamond Company's Wells.—Just commenced drilling. Situated on the Kinstler farm.

El Dorado Company's Well.—On Conver farm. Depth, five hundred and thirty-seven feet. Pumped about one barrel per day until lately. Now idle.

Pacific Oil Company's Wells.—On same farm. One is idle with tools fast; another is drilling. Below these are two wells abandoned and two new derricks, one of them belonging to the United States Petroleum Company. No progress. Next two works (one new) damaged by flood; also the relics of one burned, which belonged to the United States Petroleum Company, and said to have yielded thirty barrels per day. No signs of activity at any of these.

Marshal Well.—On Tolles farm. Been in operation about a month, and pumping thirty to forty barrels per day. Depth, three hundred and fifty feet.

Dr. Kinter's Well.—Has been pumped nine months, giving average of fifteen barrels per day. Depth, three hundred and sixty feet. Owned by Alleghany and Walnut Bend Oil Company. Another, belonging to the same company, in progress.

Drummond & Arnold Well.—On Stiner farm. Yielded freely for a year. Exhausted, and deepened without any result, and then abandoned. Depth, about five hundred feet.

M. B. Brown's Well.—In progress. Depth, four hundred and sixty feet. Another work in progress.

Rathburn, Lay & Company's Wells.—On Rhenof farm. Five in number. One sunk seven hundred and seventy-two feet. Gave no oil worth mentioning. One, five hundred and thirty-three feet, with like result. One sunk last year yields twenty-four barrels per day, on average. Other two in progress. The best of idle works yielded seven barrels daily for short time, but fell off.

Cornwall and Titus Wells. — Two in number. One pumped seven barrels per day for three or four months; the other did rather better. Both abandoned. Depth respectively, five hundred and five and six hundred feet.

Powhatan Well.—On Downie farm. Belongs to a firm at Kittaning, Pa. Sunk two hundred feet in 1861. No oil. Abandoned and now deepening to five hundred feet.

Ross Oil Company's Wells.—One is two or three years old. Before freshet yielded twelve to fifteen barrels per day. Damaged and idle at visit. Depth, four hundred and fifty feet. Another well has been five or six months in operation. Yields about ten barrels per day. Has improved since flood. Depth, four hundred and twenty-five feet. A third gave no oil and was abandoned.

Clintock Cornwall Petroleum and Mining Company.—Have just put down a well five hundred feet. Only water as yet. Four old wells and one unfinished close by.

Wood, Mc Williams and Company's Wells.—On lower Rhenof farm. Owned by a New-York interest. Put down in 1861, but neither tested. Depth, about four hundred feet. To be started anew. A third, put down same year, yielded irregularly from five to fifteen barrels per day. Depth, three hundred and twenty feet. To be also resuscitated.

Ballard Well.—Pumps twelve barrels per day. Sunk in 1861 to depth of four hundred feet. Another well of same owners never tested.

Horse Creek Eddy Well.—Has pumped as high as fifty barrels daily, with average of twenty-five barrels, past year. Opened in the spring of 1864, and now in operation. Depth, four hundred and eighty feet. Belongs to a Pittsburgh firm. Another well, owned by same, sunk in 1862 about six hundred feet. Averaged fifteen barrels per day for one year. Now idle.

Humboldt Oil Company's Wells.—A New-York concern, owning part of the Lamb farm. Eight wells completed. Three unproductive. Two pump twenty barrels each, two thirty barrels each, and one five barrels per day. The last has been going four years. Depth, three hundred and ninety feet. All productive works said to be on the increase. Three of them opened the last six months.

Tarr Homestead Oil Company's Well.—Yields from twenty to twenty-five barrels per day. Been in operation nine months, with no decrease. Depth, four hundred and sixty-two feet.

Kincaid Well.—Pumps two to three barrels daily. Sunk nearly four years ago. About a dozen works in progress close by, mostly inclining toward the uplands. One, newly opened, said to yield seven to eight barrels per day. Another testing.

Bradley Bend Wells.—Two in number and abandoned. For few weeks last summer one yielded sixty or seventy barrels per day; then gave out. Depth four hundred feet. Belongs to Carbon Oil Company of Philadelphia. The second well yielded little. A third sinking.

Sheridan Well, No. 1.—On Lay farm. Opened last February, and yields twenty to twenty-five barrels per day. Depth, three hundred and eighty-five feet. Owned by an Eastern company. Another well in progress with good appearances at two hundred feet deep.

Wheeler Wells.—On Carey (or Curry) farm. Owned by Rockford Oil Company of Philadelphia. One in operation since 1861. Ranges from two to thirty-five barrels per day, with average of twenty. Depth, three hundred and twelve feet. *No. Two* is idle. To be deepened and re-reamed out to depth of five hundred feet. *No. Three* starting. No oil as yet. *No. Four* in progress.

Baltimore Petroleum Company's Wells.—Ownership of six, on Downing farm, said to be divided between that company and a Philadelphian. Works sunk three or four years ago. One increased from five to ten barrels per day, but stopped by freshet; now pumps only one and a half. All the wells understood to have been profitable at one time; but need to be burnished up and deepened.

Alcorn (or Elkhorn) Oil Company's Wells.—On farm of same name. One begun in 1860, and sunk three hundred feet. No oil. Another put down to six hundred with like results. A third, also unproductive, is being deepened. Two others are leased to a Michigan company, who are sinking further.

Howard Well.—Sunk five hundred and twenty-five feet, and going down to seven hundred. Report of a good show, etc.

Well No. Seven.—Also on Alcorn tract. Put down in

1860, and yielded two to seven barrels daily. Engine too weak.

Caswell, Herbert & Company's Well.—On island opposite Not in operation since flood. Yielded ten to fifteen barrels a day before freshet. Now sinking deeper.

Hilands Oil Company's Well. — On Siverly farm. In progress. *Thompson Well.*—Ditto. Sunk three hundred and fifty feet. Got small quantity at one hundred and fifty feet. One abandoned well, never yielded much. Another commenced drilling.

Clark & Company's Well. — Sunk in 1861, by hand and horse-power. Never pumped, but dipped up with pail half a barrel per day. Owners now preparing to sink deeper. Belongs to Chatauqua Oil Company. Depth, four hundred feet. Another sunk in 1861 to five hundred feet, with no result.

MEM.—Siverly farm is about one mile above Oil City. Land fronting on river is held at five thousand dollars per acre.

On the Hassen farm, immediately above Oil City, are the following wells: *No. One.*—Yielded from two to eight barrels per day for a few weeks in 1860. Depth, one hundred and sixty feet. Sunk two hundred feet further with no result. *Water Works well.*—Pumped ten barrels per day for two years; now dry. This was the only old work on the farm that yielded largely. Five or six others were sunk and abandoned as unprofitable. One has continued to flow or drip about a barrel per day for three months.

Cleveland and Cherry Valley Oil Company's Wells.—Are situated on Reed Run. One pumps ten barrels per day, from depth of six hundred feet. Newly opened. Another getting ready to test, with fair prospects.

On the opposite side of the Alleghany, between Oil City and Walnut Bend, are perhaps thirty wells, a majority of them under way. I did not observe any pumping oil as I passed up the river on the west side.

In making up a recapitulation of the whole, some allowance should be made for wells actually yielding oil, but which appeared to be idle at the time, perhaps because they were pumped "by head." I have made allowance for *ten* such at the usual rate of yield on Oil Creek.

If it be desired to calculate upon the basis given underneath the *annual* product of the wells, the number of days should be set down at about three hundred and forty, to make up for the large flowing wells, which work seven days in the week, and about one-quarter of the pumping-wells, whose owners follow nature in this respect.

The number of wells denotes, not the engines which were pumping merely, but those which were pumping *oil* on the days of my visit. On the Tarr and other farms, I estimate that at least fifty more were at work, exhausting the water, making the active aggregate three hundred and seventy-five. This will probably be increased to five hundred before midsummer.

RECAPITULATION.

SECTIONS.	WELLS IN OPERATION.	TOTAL YIELD, BARRELS.	AVERAGE PER DAY.
Watson Flats, etc., ...	21	362	17.1
Miller Farm, ...	1	28	28.0
Foster & McElhenny Farms, ...	15	523	34.9
Funk & Boyd do. ...	16	648	40.5
Wash. McClintock do. ...	15	345	23.0
Hyde & Egbert do. ...	6	725	120.8
Storey do. ...	22	855	38.8
Tarr & Blood do. ...	19	411	21.6

SECTIONS.	WELLS IN OPERATION.	TOTAL YIELD, BARRELS.	AVERAGE PER DAY.
Rynd & Wid. McC. do. ...	14	147	10.5
Rouse & Buchanan do. ...	17	415	24.4
H. McClintock to Oil City,...	16	682	42.6
Oil Creek valley,............	162	5141	31.7
Cherry Run,..............	51	1972	38.7
Pithole Creek, ...	3	300	100.0
French and Sugar Creeks, ...	14	50	3.6
Lower Alleghany,...........	51	404	7.9
Upper do.	31	666	21.5
Omitted, ...	10	317	31.7
Grand totals,...........	322	8850	27.5

At this rate, the annual product of Petrolia may be set down at three million nine thousand barrels. We can very well afford to leave out of the account the odd thousands, and accept the round three millions as the amount of very sensible perspiration which has exuded from the pores of our common mother in that twenty miles square block of Pennsylvania.

But, in truth, the amount given, magnificent as it may be, is far below the grand aggregate, taking one season with another. Along the Alleghany and the lower farms on Oil Creek and Cherry Run, I estimate the proportion of wells temporarily disabled by the freshet at one-fourth of those producing in the early part of March. Again, there are large numbers which are now operated during the summer months only, and had not then got fairly under way. I think one hundred of these at least would be going by the first of May, and continue steadily during the summer months. It is true the average productiveness of these is considerably below that of the aggregate.

* Partly estimated.

But if an allowance of one-half be made for this drawback, the residue would make a material increase in the quantity. The newly opened and reöpened wells will do more, during this summer, than replace those which go out of date as non-producers. Taking all these increments into consideration, I have no doubt that the actual yield of Venango is over ten thousand barrels per day, giving as a grand annual aggregate *nearly three millions and a half of barrels.*

9

CHAPTER VII.

OIL REFINING AND REFINERIES.

The refining of petroleum, or preparing it for illuminating and lubricating purposes, is one of the new departments of industry created during the past five years, and now giving profitable employment to hundreds of men in Western Pennsylvania alone. Its object is two-fold—to free the liquid from impurities, with their offensive smell, and to render it unexplosive. In both, the most triumphant success has attended the efforts made; and rock-oil, as it affords the cheapest and most brilliant light known, is equally safe and inoffensive.

For much of this success, the country is indebted to Mr. Samuel M. Kier, of Pittsburgh, who had been giving his attention to this subject as early as 1849. He sent a quantity of it to a Philadelphia chemist to have it analyzed, and learned that if he could get a suitable lamp for burning it, the oil would make an excellent illuminator. Returning to Pittsburgh, he set other men's wits at work, and soon obtained the desideratum, Mr. Kier erecting a small refinery. From 1850 to 1855, he disposed of all the petroleum he could obtain from his own and his neighbors' *salt*-works, in which it had been discovered, selling with it the lamps used for burning it. Boring for oil exclusively had not then been thought of; but when Eveleth, Bissel, & Drake made their famous attempt, Colonel Drake visited and examined Mr. Kier's salt-wells on the Alleghany—with what result is already sufficiently known.

The first process in distillation, after pouring the crude oil into tanks in the ground, is to pump it into the stills. At Corry, three of these have a capacity of twenty-six thousand gallons each. They are made of heavy boiler-plate, capable of withstanding a very high pressure, the liquid being raised to a heat of four hundred or five hundred degrees Fahrenheit. The stills being duly charged and closed, fire is applied in the furnaces underneath, about six o'clock in the morning. By the same hour in the evening they should be emptied. As the oil evaporates, under such a powerful heat, the gas passes into a "worm" or "condenser"—a long, slender tube, immersed in a current of cold water, which causes the vapor to return to the liquid condition. On emerging from this tube, the oil has a whitish blue or bluish white color, instead of its native dark green hue.

From the condenser, it next passes into the "receiver," a large tank, out of which it is transferred, without undergoing any further change, into the "treating-tank" or "agitator." This may be of any size, provided it allow for mixing thoroughly and in due proportion the "distillate" (the name given to petroleum in that stage) with sulphuric acid, (oil of vitriol.) The quantity of acid usually assigned to thirty barrels of oil is between five and six pounds, which being poured in, the whole mass is stirred or agitated, by means of a strong current of air or iron paddles, for about ten minutes. The object of this process is to separate from the oil such foreign ingredients as dirt, tar, and other impurities, that may still have mingled with it. These settle down and rest upon the concave bottom of the vessel by their own greater specific gravity, and are taken away separately. Their color is dark purple or nearly black.

The oil is next washed with clean water, and agitated afresh for a period of fifteen or twenty minutes, this process being repeated several times, so as to remove from it every particle of the *acid.* After washing, it is treated to a dose of alkali, (usually soda,) in the proportion of five gallons to thirty barrels, whereupon it is submitted to the "hydrometrical test," to be noticed afterward. The alkali gives it brilliancy, and removes every particle of acid that may have remained in it. Next, it is drawn into the "bleachers" or "settling-tanks," which are large, shallow wooden tubs. The liquid has now a whitish or bluish-white color. Measured by the hydrometer, this instrument is found to rest at the point marked forty-seven degrees on the scale, sinking to the level of a higher figure in proportion to the lightness of the oil. It's range varies from fifteen to eighty degrees. The coal-oil is not so white as that distilled from petroleum, which is more of a straw-color than the former.

From the settling-tank, it is then drawn off into barrels of from forty to forty-five gallons each. These are made of the best white oak, their insides being carefully lined with glue or soluble glass, previous to being filled. If the barrels be sent back for re-filling, the same process has again to be gone through with; it being judged unsafe, by means of a crack in the coating, to leave a particle of the wood exposed to the insinuating action of petroleum. Besides the loss of material in such a case, there is the much greater danger of the liquid, after oozing through, taking fire and exciting a general conflagration. Before re-gluing or re-glazing, the old coat has to be melted and drawn off by means of a jet of steam.

After distillation, the first liquor that comes off is *naphtha* or *benzine*, a very light, volatile, and inflammable substance, its hydrometrical test varying from sixty-five to seventy-five degrees. When the discharge coming from the condenser descends to sixty or sixty-two, the naphtha is cut off and let run to oil. If cut off at sixty-five or seventy, the oil will be very light, rather inflammable, and insufficient to stand the fire-test. In different states, various tests as to the quality of the liquid in this respect have been established, the instrument used being termed a "pyrometer" or "fire-measure." The standard in New-York is one hundred and ten degrees, but in Pennsylvania and Ohio one hundred. The pyrometer is a little brass vessel, the bottom of which contains a small quantity of water, heated by applying a gas-lamp. Immediately above this water is placed a small cup containing refined oil, which is heated equally with the water. The "test" means the point at which vapor arising from the petroleum, after the application of heat, will *ignite*, when a lighted match is held above it. If it takes fire at a low figure in the scale, there is danger of such oil, when placed in a lamp, igniting and causing the lamp to burst, since a certain portion of heat is communicated to it all through from the flame above. If the gas blow out the light, it is termed the "vapor-test;" if it ignite, it is the regular fire-test. The difference between these is not apt to be quite five degrees. The point of ignition is between one hundred and fifteen and one hundred and twenty degrees in the best quality of petroleum. It may be observed, however, that, in consequence of its more rapid heating by some persons than by others, this standard will vary somewhat; the difference between the figures, in testing the same sample, running from one to five degrees, according as the person engaged applies the heat rapidly or slowly. The more time there is allowed, the article will appear to better advantage.

There are three grades of refined oil—"the prime white," "the standard white" or "light straw," and the "straw-colored." The last-named usually stands a fire-test of one hundred and fifteen degrees.

The next run from the stills, after benzine and oil, is called *paraffine*—a whitish, wax-like, inflammable substance, which is used in making the best candles, in mixing with wax, and occasionally is manufactured into sweetmeats! The proportion of this substance to the pure petroleum varies according to the season, being greater in winter than in summer, when some of it appears to mingle with the oil. Out of one hundred barrels of distillate, from three to five of paraffine will usually be obtained. All refiners, however, do not separate it, some letting it remain with the liquid. Elsewhere I have noticed the troubles arising from the collecting of this ingredient on the inside of the tubing, the pump-rods, and even the well, stopping the veins and preventing the passage of oil upward.

The residuum coming from the condenser is *tar*, the ratio of which to all other substances is under one per cent. It is commonly used on the ground as fuel; but a new use for it has been discovered in some places where refineries have been established, namely, to lubricate such heavy articles of machinery as cog-wheels.

The fine lubricating oil obtained on French and Sugar Creeks, and for a short distance above the mouth of the latter on the Alleghany, is not always submitted to the refining processes. Such portions of it as are used on locomotives, stationary engines, and most kinds of heavy machinery, pass directly from the tank to the operative's flask; but it is judged best to refine such oils as may be required on the more delicate kinds, as cotton, silk, and woollen machinery. The modes of treating it do not vary materially from those of refining common illuminating oil.

In the Corry works, after going through the several processes, it is placed in a room which is a mammoth refrigerator, the temperature being reduced to a very low degree by salt and ice. In a short time, the oil becomes a thick slush, in which condition it is put into strong canvas-bags and subjected to a powerful pressure, by which the pure oil is forced through the coarse cloth, leaving the paraffine inside, which forms in thin, hard cakes, of a gray or light-brown color. This substance readily separates into thin flakes, somewhat resembling the scales of a fish. In this condition the paraffine is sent to Boston, where it is refined to the pure white article of commerce.

The quantity of merchantable products of all kinds—refined oil, naphtha, paraffine, and tar or residuum—obtained by distillation, varies according to the quality of the crude oil, and to some slight extent according to the season. At the Downer works they reckon upon eighty-five to ninety per cent on the average. Mr. Sommers of Jersey City assures me that he has distilled as much as ninety-five per cent out of some illuminating oils; but this is an unusually high figure. There is commonly a little more loss in refining oils for exportation than for consumption in the home market.

As to the proportions of *each,* there is much diversity, these varying with the natural quality of the crude article, and with the purpose and skill of the manufacturer, so that it is difficult to give an estimate which will apply to all. Mr. Sommers states that when they make it for exportation, the fire-test being one hundred and fifteen, they get about seventy-five per cent of refined petroleum and about fifteen per cent of benzine. With a fire-test of one hundred degrees, which is considered perfectly safe, they get from eighty to eighty-two per cent of refined oil, and about eight per cent of naphtha. The latter ordinarily sells at about half the price per gallon of the former.

One of the products of Heinrich & Sommers's refinery in Jersey city is *gasoline*—a liquid which bears about the same relation to benzine as the latter does to refined oil. It begins to come off at a heat of eighty-five degrees Fahrenheit, and is cut off when the heat rises to one hundred and fifty degrees. Between one hundred and fifty and two hundred and twelve degrees, (the boiling point,) benzine or naphtha comes off. From this point, to get all the petroleum, the heat is raised to about four hundred and fifty degrees, beyond which it is scarcely ever necessary to go. Gasoline is used in some establishments for the manufacture of common illuminating gas, which can be made from it more cheaply than from coal. It has been introduced into the Springfield armory, among other concerns. The government charges a tax of five per cent *ad valorem* on this, as on refined lubricating oil, instead of the ordinary twenty cents per gallon.

The proportion of crude to refined oil exported, Mr. Sommers estimates at from one-third to one-fourth. More or less of the former is sent to Great Britain, France, and Germany, where chemicals as well as labor can be had cheaper. To all other countries petroleum is shipped in its refined state only.

There is no single centre of this business, it being carried on all over the country, from the sea-board cities to Corry, Pittsburgh, and numerous other points in the West. The largest concern of the kind in Petrolia proper is the Downer works at Corry, the premises comprising half a dozen acres. The number of men employed usually approaches two hundred, and the capacity of the works is eighteen hundred barrels per week. The buildings are of brick and made fire-proof throughout. On one occasion a still burst and its contents took fire; but as each of the six furnaces occupies a separate apartment, the flame was extinguished by the application of steam from the adjoining boilers. The whole arrangements about this concern are equally perfect, showing a regard to order and neatness as well as safety.

Next in capacity to this establishment is the Humboldt refinery at Plumer, the area inclosed being nearly twenty-five acres, and the works so planned as to take advantage of the natural descent of the ground, in the passage of oil from one set of vessels to another, thus dispensing with the use of artificial power; they are also so far separated that, in the event of one building taking fire, it would not be communicated to any of the others. The proprietors are Messrs. Ludovici, Brothers, who are natives of Germany, and gentlemen of rare intelligence and urbanity. The author of a pamphlet entitled, "All about Petroleum," has fallen into some amusing errors concerning this establishment, it is believed, from not having been within some miles of the concern which he attempts to describe with his usual grandiloquence. He speaks of "a color produced from the residuum of the petroleum" at that establishment, as being "a bright and fixed cerulean blue, or perhaps a shade darker, and called the Humboldt color." Its proper name is *analine.* One of the proprietors assures me that, at the time when this statement appeared, they had not even thought of making such a color; but since that time, inquiries and orders had come to them in such numbers that they concluded to engage in it. Further, the writer goes on to say, that "the discoverers are German chemists, who do not speak, if they understand, English." The Mr. Ludovici whom I saw speaks as good English as is found in the pamphlet! Again: "No stranger is allowed to enter their works, except by special permission." Nothing could be more incorrect than this statement, as I wandered through them and made inquiries, without so much as one of the employés asking a question, much less refusing permission to go further. Other persons did the same thing while I was present; and this was no newly adopted arrangement. So much for the accounts of a writer who, from a distance of miles, undertakes to describe "All about Petroleum," manifestly taking less pains to arrive at the truth than to depict, in glowing "colors," a great and important interest, whose defects, however, he is equally careful to conceal.

The number of stills in the Humboldt works is twenty, and their capacity is one thousand barrels per week. The oil refined there is for the most part exported to Europe, where it has an established reputation. The proprietors are organizing under a charter of incorporation. From Tarr farm, three miles distant, the oil raised is forced, by a powerful pumping apparatus, over the hills to Plumer, and, after refining, is thence taken by wagon to the Alleghany River landing or to Titusville. Saw-mills, barrel-factories, etc., are on the premises.

Messrs. Warren, Brothers', refinery is in the same village, and contains sixteen stills, with a capacity of nine hundred barrels per week. A powerful Worthington pump forces the refined oil to the summit of the ridge east of the works, about two hundred feet in perpendicular height, whence it descends by the force of gravity to the Alleghany.

At Titusville are four or five refineries, mostly of small capacity, the principal being the Bunker Hill works; probably as many as twenty more of the same character are scattered along Oil Creek down to its mouth. Near Petroleum Centre are six of these, the largest having four thirty-barrel stills; three are on the Storey farm; about half a dozen are on the flats immediately above Oil City, the largest being the Union works, with four stills, and now owned by Mr. Truair. In the lower part of that place, the Oil City Petroleum and Refining Company own a refinery having four stills, with a capacity of three hundred barrels per week, and a second, of two hundred barrels, on Holliday Run. Messrs. W. H. Lay & Co. have another, with three stills, and a capacity of two hundred barrels. On Cherry Run, the Messrs. Orr's refinery will turn out one hundred and fifty barrels weekly. At Reno Station, two miles below Oil City, are three refineries; about as far above Franklin, another is about to commence work. One mile above Franklin, on French Creek, a neat little concern has been in successful operation for some time.

On the flats above Oil City, only one of these establishments was actively at work when I visited that locality; and taking all the works along Oil Creek, as many as one half were standing idle. The others were moving along under easy sail, so that it is probable three times the quantity of refined oil could be produced that was then being turned out. As a whole, I have no doubt the business has been profitable, and will so continue to be, in spite of the oppressively heavy taxes imposed on both crude and refined petroleum by the general government. It was represented, however, that, in consequence of these, some works had suspended operations, and might come to a full stop. If this imposition, made by Uncle Samuel, be traceable to the impositions made upon many of Uncle Samuel's *family*, through misrepresentation as to the resources of Petrolia, who shall say that the retribution was not deserved?

The advantages and disadvantages of establishing refineries near the oil regions are many, and pretty nearly equally balanced. By being on the ground, one is enabled to take immediate advantage of every turn of the market in making purchases. There is also a saving in transportation; though this is partly offset by the freight on chemicals, which have to be sent on from the East. Probably the greatest benefit is in knowing and being personally known by managers at the wells. On the other hand, real estate, fuel, labor, etc., are from fifty to one hundred per cent higher in Petrolia than elsewhere; and until "order reigns" there, must so continue to be. With cheap transportation by river, and thence by railroad, with an abundantly supplied labor market, and coal at a minimum, Pittsburgh is thought by many to be the best point for refining petroleum; but with the requisite capital and skill, a small refinery may be run successfully at almost any point furnishing cheap fuel, cheap transportation, and a market for the article. No other form of industry is likely to be less localized than this. The principal matter to be considered is, whether the number of refineries is not already in excess of the supply of crude petroleum

CHAPTER VIII.

HOW STRANGERS ARE TAKEN IN.

"If we succeed, we shall make half a million; if we don't, we can lose only the five hundred dollars," observed an oil-prince to me one evening, after relating the success of a pecuniary transaction in which he and a few others had engaged. The prince aforesaid is by no means an unworthy citizen. On the contrary, as he is rotund and good-natured, so he is affable and obliging. I am not certain but that he is even public-spirited—at least as much so as any body else in the same locality. He was concerned in no operation which society would term nefarious, much less stamp as infamous; but with all this, he, a dull-looking, phlegmatic Pennsylvanian, was setting his snares for catching smart, shrewd, keen, sagacious Bostonians, New-Yorkers, and Philadelphians, of the Wall-street type. And as he was ready enough, without solicitation, to unfold his plans, I have thought proper to open this chapter with an allusion to them.

The mode of operating was substantially this: He and his associates had purchased a considerable tract of land at the head of ——— Run, on which the sum of five hundred dollars only had to be paid down; the residue to be paid at a date sufficiently far distant to allow time for making experiments. At the same time, the associates had made arrangements with another person to sink a well on an acre-lot—they contributing with the real estate one-tenth of the capital, for which they were to receive one-eighth of the oil obtained. The locality being at some miles' distance from any paying well, it is easy to see that this bargain was perfectly satisfactory to the purchasers, as the risk of sinking was nearly all transferred to the well-digger. If he succeeded in getting oil, one acre of the remaining property would sell for as much as the whole tract had cost; if not, they had only to throw up the contract and forfeit the five hundred dollars. Hence it was with a smile of triumph that he made the remark: "If we succeed, we shall make half a million by the operation."

By transactions somewhat similar, nearly every foot of land within fifteen miles of Oil Creek has passed into the hands of speculators, who are "operating for a rise" by arts and appliances known only to the initiated. Some of these are not dishonorable, in the ordinary sense of the term; others are bald swindles which call for exposure.

Lands are not offered for sale by middle-men until a well has been struck on or near the premises. No matter how remote these may be, the "good show" of petroleum is inevitably followed by the erection of one or more lines of telegraph. Every day's yield is carefully noted and registered and telegraphed to the great cities, as also to the principal points in Petrolia where strangers are accustomed to congregate. This operation continues as long as the well keeps on the increase, and even after it comes to a stand-still. The philosophy of laying down telegraphic wires, while common highways have hardly been thought of, is explained by the maxim: "Strike while the iron is hot." In other words, sell your interest in a well before its productiveness has begun to fall off, which is morally certain to take place within a calendar month after discovery. The market value of a good well on Oil Creek or its tributaries is usually calculated at so much for every barrel of oil yielded in one day. Last winter the price was as high as five thousand dollars per barrel; since then, it has fallen off to three thousand dollars. Thus a forty-barrel well is supposed to be intrinsically worth one hundred and twenty thousand dollars. On French or Sugar Creek, where the oil is of a much better quality, the price per barrel, of course, is greater.

Now it is of the utmost consequence that the measurement be made and widely published while the well is in the hey-day of its prime, before its energies have begun to relax and its discharges to be less copious. At five thousand dollars per barrel, a newly-tapped source giving one hundred barrels per day might have realized half a million last February or March; while by the middle of April it would probably have declined to fifty or sixty barrels, worth only three thousand dollars each, or say one hundred and seventy-five thousand dollars. The reader will readily comprehend the zeal manifested for "progress" in the direction of the magnetic wire throughout Petrolia; while such every-day matters as common roads are utterly neglected.

To coöperate with the telegraph, the several entrances into the oil region are commonly garrisoned by a corps of veterans, who have a direct interest in magnifying the powers and resources of the country. With the prosperity of Oil-dom proper, all the avenues leading thither also prosper; every hotel-keeper, every land-agent, every merchant, every owner of real estate, prosper. Demetrius, the silversmith, calling his workmen together and explaining to them, "Sirs, ye know that it is by these things we have our wealth. . . . Great is Diana of the Ephesians!" explains sufficiently the influences which are often brought to bear upon the stranger, as he passes through Corry, Titusville, Franklin, or Oil City. "Have you heard of the two hundred-barrel well struck on Big Pithole yesterday? It is a fact, sir. I saw a gentleman just come from it, and he assured me that he 'timed' it with his watch for a quarter of an hour, and it gave one hundred gallons good in that time. I have no doubt of his word, sir. Pithole is going to be developed this summer. The best judges say it is every bit as good oil territory as the creek." To which the company respond affirmatively, some of them perhaps announcing still greater developments in the same line.

For the group present consists largely of interested witnesses. In those hotels and boarding-houses on the outskirts of Petrolia are platoons of men eager to "make an honest penny" by selling or leasing lands, the right of refusal, oil stocks, interests in wells, etc.—their profits or commissions depending on the result, according as they may succeed in making a favorable impression or otherwise on the stranger. There are also congregated the agents of Eastern manufacturers, who strive to push their fabrics as extensively as possible into the country. There, too, assemble daily the oil-princes who have made their fortunes in the business, and have either retired or kept extending the circle of their operations, until their names appearing as managers of any new enterprise secure it all the pecuniary support required. The sight of a living, moving, talking *millionaire*—perhaps only a teamster three years ago—is too dazzling to most minds to be gazed upon without turning the head. As he paces the floor, and proceeds to narrate his experience in the oil regions—how he came hither with only five dollars in his pocket, and how he stuck to it till he found himself the owner of a controlling interest in eleven paying wells, sandwiching advices among the observations—it is difficult to escape catching the contagion. "I never knew a man that stuck on to the oil business but what succeeded, and got rich at last," is a clincher to the stranger with one hundred thousand dollars in money or credit in his wallet.

Mr. Secretary, we will suppose, has just arrived at one of these caravansaries, and become deeply interested in conversation with an oil-prince or agent, who depicts in glowing terms the virtues of a certain piece of property, and the productiveness of a well, in which he has an interest for sale, after having made money enough, etc. As the agent or confidential adviser of a strong company at the East, he decides to visit that well the next day to see for himself, and if the facts turn out as represented, either to purchase an interest in it, or "territory" as close by as possible. He will recommend a purchase, if he does not himself buy; for oil is now down in price, and certain to advance with the opening of the spring business, so that the purchaser of a large interest is likely to make a good thing out of it before thirty days. But he is too old a bird to be caught with chaff; he will place implicit belief in nobody's word; he has seen too much of the world for *that.* He will go and examine for himself what the well is actually doing, "timing" its yield, and making all the other necessary inquiries.

Now here comes in the great beauty, the amazing utility of that marvel of the nineteenth century, the telegraph. For while Mr. Secretary is admiring his black boot-legs, or calculating his profits on the prospective purchase, the affable Petrolian who gave him the information and offered to accompany him to the well, has quietly sent a dispatch to the manager with this purport: "Secretary and I will be with you at eleven o'clock to-morrow. Have every thing in apple-pie order. Yours, PETER O'LEUM."

Peter, you must know, is the most disinterested fellow in the world, and clever withal. He cracks jokes at his own or other people's expense. On their way he is voluble in explaining every thing—the phenomenon of so many wells standing idle and the like. They were put down just before the great fall in the price of oil, four years ago, and then abandoned; the company had not means enough to finish, although the well gave a first-rate show; the great freshet—that convenient scapegoat—swept over the bottom, and managers have not yet got their new machinery on the ground; such a well paid for itself five times over, and can afford to rest awhile; yonder is the Great Geyser, which literally set the river on fire and burned for three days and nights. With such thrilling reminiscences of the past, the stranger at length finds himself confronting the object of his search—the two hundred barrel well.

And it *is* a grand sight to behold!. For, even after reducing its reported yield fifty per cent, the spectacle of one pouring forth of its own accord one hundred barrels every twenty-four hours is sufficient to make the beholder's eyes glisten, his teeth water, and his brain grow dizzy, unless his mental composition be different from that of most men. That *moderate* quantity in Petrolia, at the low prices lately ruling, represents an income of five hundred dollars per day, fifteen thousand dollars per month, or one hundred and eighty thousand dollars for the whole year. True, it is far from being equal to the wealth of a Tarr, a Blood, a Hyde, a McClintock, a Culver, a Sherman, a Noble, a Delamater, an Allen, a Funk, a Downer, or many another oil-prince; still it is a snug little competency with which an aged pair might think of retiring, without feeling greatly distressed as to the means of support in this world. The very fumes of the gas have an exhilarating, if not intoxicating, effect on one's brain.

Who cares, under circumstances where he is likely to be pecuniarily concerned, about the mystery of the genesis of petroleum? Of what consequence to the purchaser whether the drill has struck the aorta of a half-petrified whale, which is making its last and greatest spout, as it was wont to do in the Greenland seas; or whether the greasy liquid has been distilled from coal, or is a new chemical combination going on in the world's basement-story? Topsy's explanation, that it "wasn't made, but growed," is as good as any when one is seriously thinking of making an investment. "Well, sir," observes Mr. P. O'L., after allowing the spectacle time to make its own impression, "you see there is no mistake about that well. I have timed her repeatedly, and found her to be a little over two hundred barrels a day. The engineer says she yields better at night than during the day, and in winter quite as good as during the summer, on account of the paraffine which collects in the warm weather, you know. That well would net you a good half-million the first year, and in twenty years would make a man as rich as John Jacob Astor."

"Figures never lie;" "facts are stubborn things," and all that; but with any number of such proverbs and wise saws, it is almost refreshing to watch how many shrewd, sharp, intelligent Eastern financiers, who feel themselves competent to buy and sell all creation, can themselves be bought and sold and delivered by Petrolian speculators, through the simple agency of facts and figures.

Perhaps the disinterested Peter is agent for an old and established concern—a fifty-barrel well, we shall suppose —that has a history of which any institution might justly feel proud. It has already enriched half a dozen; and the lucky owner of a one-fourth interest in it, having made money enough, is desirous of leaving the country, and enjoying his dignity with ease and plenty elsewhere. Peter induces the stranger to fix the day and hour for a visit, whereupon the secret is duly whispered over the wires. Now, it happens that the aforesaid well has, for a year or more, been pumping "by head," that is, two or three hours per day, or as often as the butter collected on the churned milk below, thus saving both fuel and labor. The men engaged about it have received instructions as to how they shall answer questions; and so, every thing being in readiness, the curtain rises. Mr. Secretary and Peter step out on the stage and examine the works. They repair to the tank, climbing the greasy ladder which leads up to it, both manifesting the utmost unconcern about purity of fingers or cleanness of coat-skirt. The fact is, in twenty-four hours the stranger has become almost a Petrolian; and Peter is careful to flatter his vanity, by congratulating him on the rapidity of his naturalization. Peter's mouth yields more lubricating oil than the best well on French Creek. "A big thing that, I tell you," he exclaims. "It is," dryly responds Secretary; "but I guess the yield's not over thirty-five barrels a day." The proprietor here steps forward, and observes: "You see that tank, sir. It is ten feet deep, contains eight hundred barrels, when filled, and is now half full. (Measures.) Well, sir, that was pumped since last Monday week, just eight days." In verification of this he calls for the engineer, who confirms all he had said, except that it lacks an hour or two of being *quite* eight days. "Well, I knew," proceeds the other, "that I had not gone beyond the truth, at any rate. She's one of the best and steadiest wells on the creek; and if I didn't want to get out of the business and go home to my family, I would never think of parting with her." What more could Mr. Incredulity himself ask than this? To the testimony of three witnesses is added ample verification by his own eye, he having timed the yield by his elegant repeater. He inquires the price, the lowest cash-down price, at which the owner is ready to sell his interest, and closes the bargain then and there.

> "Which certainly seemed wondrous cheap,
> And for the money quite a heap,
> As every man would buy with cash and sense."

MEM. NO. ONE.—If Mr. Secretary had been there three days ago, he might have learned that a purchase of fifty barrels of oil had been made from a neighbor, the same being conveyed on the sly into the tank of that "very steady well."

MEM. NO. TWO.—If he had some means of gauging its contents, he might discover that seventy per cent consists of *salt water*, which has settled to the bottom by virtue of its greater specific gravity. Ordinarily, this is permitted to escape by a stop-cock, inserted near the bottom; but this discharge has been stopped, to prepare for such distinguished visitors a suitable reception.

MEM. NO. THREE.—If he should remain on the ground two hours, he would be certain to find the engine breaking down, so that it could not pump another drop. By some lucky arrangement a bolt would get loose or fuel would run short, just in the nick of time, the oil having been completely exhausted from the well! Whereupon the engine would get a blessing (in disguise) from some of the men in charge of the works, accompanied by the frank declaration, that if they had only got a good machine, instead of that rickety concern, the well would have yielded sixty or seventy barrels a day—perhaps one hundred.

The purchaser is a believer in improvements, and makes up his mind to replace that inefficient engine by a good one, which he proceeds to recommend or order as soon as possible.

That evening he retires to rest in a delicious state of self-complacency at his day's work, and full of great expectations for the future. The seller sticks his tongue in his cheek and keeps it there. Peter has managed to clear a thousand dollars by the operation, besides gaining the reputation of being a deuced smart fellow.

I assert that this is a part, and but a part, of the system of swindling carried on in the oil regions—a system which has been reduced to both a science and an art. It is exquisite, magnificent, stupendous, brilliantly successful. It is the key to that certain wealth which, we are told, finally awaits all who persevere in Petrolia. The employés of individuals and companies have addressed me more than once after this fashion: "I don't want my name to appear in it; but I have been instructed to say this well yields twenty-five barrels a day. She really gives between twelve and fifteen, when she is pumped regularly; but she has not been for the past three months. You may set down her actual yield, one day with another, at between eight and ten barrels." On the lower Alleghany, I heard of a well lately sold, as giving thirty barrels a day; the actual yield was from three to four! Nor are such deceptions and frauds infrequent; there is reason to believe that one-half of the transactions at buying and selling such works, or interests in them, over the whole region are more or less tinctured with fraud and falsehood.

Some time since an anecdote went the rounds of the papers, about a well-owner near Franklin having discovered the perpetual motion in the shape of a secret tank and tube, which conveyed oil from his cistern back to the well, from which it continued to be pumped and re-pumped, as evidence that the property was a first-class concern. The man in charge of it denied to me that the statement had any foundation in fact; and I here give him the benefit of the denial, which, however, was made before any charge or insinuation was preferred on my part. The well he did not represent as doing more than pay working expenses.

Some years ago a firm engaged in sinking a well on their premises, situated on an important tributary of Oil Creek. It reached the depth of seven hundred feet, "with an excellent show" of oil; indeed, it pumped a quantity in presence of a crowd of excited spectators, and thus contributed greatly to the "development" of that "territory." But somehow matters went wrong; and, in spite of favorable appearances, the owners let that fount of wealth lie still. One of them informed me, "she was never fairly tested," which is, I believe, the case with all unprofitable works of the sort. "In wine is truth," says the proverb; and from one who was approaching the "highfalutin" state I learned that the proprietors had judiciously let down the contents of a few barrels, by night, into the orifice, to the great appreciation of values in that remote neighborhood.

There is a story told about a smart widow who resorted to the same means for creating good "surface indications," enabling her to sell her farm at a high figure. Nature was, however, smarter than she; for the first well put down proved highly productive—giving one hundred barrels a day, in fact. Unfortunately, all outsiders who thus get sold, do not fare quite so well in the long run.

A still more bold and successful mode of misrepresentation and swindling is by the employment of mercenary professors, pamphleteers, correspondents, etc., to glorify all that appertains to Petrolia, excepting the mud and perhaps the hotels, which can be roundly abused without any detriment to the oil interest. Not that every penny-a-liner is necessarily *particeps criminis;* for some have told the truth to the best of their knowledge, and only lacked time, patience, and perhaps means to remain and make their inquiries more thorough. Still, there is as little doubt that others, of the sensation class, have deliberately lent themselves to mystify the outside public, by dilating largely on the productiveness of certain first-class wells, while scarcely squinting at the hundred times as many which had been sunk and yielded nothing; by recording carefully the careers of individual oil princes, while passing by the thousands of people who had invested largely

10

and got back nothing. Any such one-sided version is, if made with the design of diverting more men and more money to the oil region, or appreciating the value of certain kinds of property, *a crime*, which should expose the guilty person to the severest condemnation, if there is no other mode of bringing him to a sense of his wrong. Even the careless collection for publication of erroneous statistics bearing on the subject, is worthy of censure, though the person's motive may not have been base. A gossipy volume on the subject, just issued from the press, gives a "rough recapitulation" of the wells on Oil Creek actually put down at four hundred and eighty, of which one hundred and eighty-nine are producing wells. In truth, the whole number of wells on that creek, within four miles of Oil City, is above five hundred, and on the creek to Titusville, nearly two thousand. A *rough* enough recapitulation, in all conscience, but misleading the public as to the proportion of producing to non-producing wells. In another part of the same volume the writer admits that the ratio of paying to non-paying wells is not more than one in ten.*

The fact is, that by relying principally upon official reports or the statements of interested parties, newspaper correspondents are as likely to be duped as other people. It requires not a day or two, but weeks of laborious effort, to reach the THIRD SAND-ROCK OF PETROLIA, before

* The same looseness in the use of figures is noticeable in all parts of this scrap-book. In one chapter, the hills near Oil Creek are described as being from two hundred to one thousand feet high. There is not *one* height in the county five hundred feet above the Alleghany! Elsewhere we read of derricks being one hundred feet high, and the like! Either avoid figures altogether or use them accurately.

accomplishing which the visitor must gain the confidence of all engaged about the works—drillers, engineers, blacksmiths, etc., as well as officials and proprietors. The impressions which I formed during the first week, although actively at work every day, were crude and incorrect in sundry particulars; how much more if I had been most of the time at the finger-ends of agents and speculators, who, in return for courtesies shown, expected to receive "first-rate notices" in my next letter.

What I blame in most persons who have attempted to write descriptions of the oil regions is this: The *prizes* in Petrolian lotteries are dramaticised, historicised, financicised, statisticised to the point of weariness, if not disgust; while scarcely so much as a passing allusion is paid to the tenfold more numerous *blanks*, which are glossed over without eliciting a warning note or word of advice. By thus concealing one side of the truth, a man communicates to the public essential falsehood. He becomes an agent in the work of deception and roguery. "The whole truth" is as essential in giving testimony as "nothing but the truth." The Sherman, the Noble, the Empire, the Philips, the Mountain, the Reed, and other wells are familiar in men's lips everywhere. Not an adult male, from Maine to California, but has heard of their wonderful performances, dreamed of the fabulous wealth poured by them into individuals' and companies' laps. But the reading public have *not* learned (the fact having been studiously concealed) that nine out of every ten, if not nineteen out of every twenty, wells sunk since 1859 have not paid their first cost; and that a portion of the residue have done little more than return cost and operating expenses. The public have not been informed that the best wells give out in time—ordinarily in eighteen months; and that the copious supplies poured forth when first striking oil are morally certain to diminish fifty per cent within thirty days, and seventy-five per cent before six months. They have not been told that the princely fortunes acquired have been gained less by the *bona-fide* yield of petroleum than by speculation, which involves more or less of misrepresentation and fraud practised upon strangers. The credit side of the account is glowingly depicted—its few millionaires ablaze with diamonds, which, unfortunately, make their original lack of culture only the more ridiculous; the debit side, representing hundreds of thousands who have invested nearly all their surplus means, in a majority of cases never to be returned, is covered with a fly-leaf, in order that fresh batches of pilgrims may prostrate themselves before the wheels of this Juggernaut.

Nor are the men who write for our public prints alone to blame. The fact is, the popular taste gives character to the print. Many men can no more do without their daily sensation than topers without their morning "bitters." They would rather be gulled, cheated, victimized by an editor than take his paper if it be *dull*. They have their reward! The morning paper which stands at the head of American sensation journals led off in the excitement about Petrolia. Thousands of its readers have seen their tills emptied in consequence; but what of this, as long as they enjoyed the luxury of a fresh sensation?

There is also on the part of others a feeling of what may be termed *shameless shame*, sometimes a species of malicious pleasure, which, while refusing to confess having been victimized, enjoys the comfort of being in company. Misery derives consolation from seeing others as miserable as itself. Instances have not been wanting of persons who, having lost all in wild adventures, chuckled at their neighbors being also heavy losers. Hence the returned Petrolian, however sorely disappointed, often prefers placing his finger on his lip to making disclosures which might cause him to be laughed at. Indeed, this disposition to enjoy others' misfortunes is only one manifestation of the wrong complained of.

But while the public receive one-sided versions of the country and its production, rash persons will always be found in abundance to rush to the "diggings," or to invest their surplus means in new or old oil-wells, regardless of the excessive royalties charged, the cost of land, labor, and materials. These may be double or quadruple what they ought to be, while petroleum may have sunk in price from one-third to one-half its former figure, with a new government duty weighing heavily upon it. What cares Recklessness, young, daring, impetuous, believing in chance, and blinded by the glitter of Peter O'Leum's equipage and jewelry? He is not going to calculate probabilities, like some old fogy, but rush in and trust to luck. If he cuts his wisdom-teeth in Petrolia and comes back a more prudent man, he has scarcely just cause to regret having parted with all his loose change to master such a lesson.

One mode by which shrewd operators contrive to fleece outside agents and others is this: On lands lying contiguous to productive wells, called, in the jargon of the country, "good oil territory," they erect derricks and begin to bore, not with the direct object of reaching, pumping, and selling petroleum, but to part with *interests* in the work, as it progresses, or after completion, at extravagant figures. The stranger comes along, observes the operations, asks the usual questions, and then reasons thus: "After all, *I* know nothing about 'surface indications,' or other signs by which experts can detect or infer the existence of oil. But yonder is the Great Geyser, yielding two hundred barrels per day, and here are experienced operators, who have invested their means in an enterprise promising, as they believe, an abundant return. They expect to strike oil before three days, and I can readily believe them, seeing they have put their own money into the undertaking. They ask ten thousand dollars for a quarter-interest in the well, and if they only succeed, what is a paltry ten thousand to a company like ours? Twenty-five barrels a day, at six dollars per barrel, [the estimated 'unit of measurement' is a yield of one hundred barrels per diem,] would repay the cost in little more than eleven weeks, and all beyond will be profit. The best thing I can do is to accept the offer and close the bargain forthwith, since, after striking oil, they would not sell the interest for ten times as much." So he buys. The same operation is repeated by others visiting the ground, until perhaps scarcely a fraction of the original interest remains in the hands of the operators, who have meanwhile pocketed several times the amount of their actual outlay. Oil or no oil is of the slightest consequence to them; they can pull up stakes and renew the proceedings elsewhere.

By such methods hundreds of thousands of dollars change hands every year in the oil regions; and a considerable body of industrious operators, knowing precisely as much of the oil-veins as do the strangers thus sold, make fortunes in a few years. Nor is it necessary to resort to very many or very glaring falsehoods for this purpose, because the purchasers, as a rule, have been intoxicated before entering the valleys. Indeed a little bravado something like, "We don't know whether we shall certainly get oil or not, mister. We must take our chances, and feel satisfied with the prospect," is often all that is needed as condiment to the more solid facts and figures derivable from the history of other wells.

In describing the swindling operations or sharp practice resorted to in the oil regions, it would be a glaring oversight to omit mention of those perpetrated by engineers (civil and uncivil) who have visited that country, bought lands, leased lands, or secured the *refusal* of lands for a time, and then gone East or West to organize companies for "developing" their "territory." Mr. C. E. visits his sweet native village or city, taking care to announce his advent in advance, and giving rumor sufficient hints that he has made his "pile," to insure a sufficiently favorable reception. This impression he confirms rather by innuendoes than by positive assertions such as might bring him into too close contact with internal-revenue agents; much less does he explain the modes by which he has accomplished his great work, lest it might offend the more fastidiously honest. He has maps of the country to exhibit for nothing, the places where he proposes to operate being marked in blue and gold. He explains gratuitously how beautifully they line in to A. B.'s "territory," from which hundreds of barrels of oil are taken every day, and of which a one-sixteenth interest sold for thirty thousand dollars the other week; how trifling is the cost compared with "territory" purchased by other companies in the same neighborhood; how he is ready to organize a stock company, with two hundred and fifty thousand dollars capital, and then put down twenty wells the ensuing summer. He has taken care to procure the names of two or three prominent men as president and directors—men who have probably no more practical knowledge of the business than they have of whale-fishing, but will leave the entire management in his own hands. The shares are put at figures sufficiently low to be taken by every housemaid in the place; and besides this, it is announced that only a nominal sum will be called for upon each share taken. The project takes, of course; for men, whether in cities or villages, are gullible beings. The scheme is largely advertised in the local papers, and receives the usual "first-rate notices," as what scheme with money in it will not? Shares are freely distributed among friends of the management and others, whose countenance may be judged necessary to the success of the project. [The outside world, and even ordinary stockholders, know nothing about the science of "watering" stocks.] In the mean time, C. E., Esq., who bought the tract at two hundred and fifty dollars per acre, or the right of refusal for fifteen hundred dollars, has disposed of it to the new company at twenty-five hundred dollars per acre, besides retaining for himself a comfortable berth as an officer on the ground. He *ought* to have cleared one hundred thousand dollars by the transaction, or he has sold the company "dirt cheap." Should any of the other officials visit the property, they may find it to consist of a narrow strip of bottom-land or ravine, the residue being mountain, from which the prospect of every thing *except* oil is entrancing.

If this company-maker be not a civil engineer, but a retired oil-prince, who has "made money enough," and has no ambition to operate further, except with the laudable intention of benefiting his less affluent neighbors—he having a dash of benevolence in his mental composition—his success in procuring subscriptions may with certainty be predicted. For he has broad acres, a costly establishment, a gorgeous equipage, a bank-book, a bejeweled wife, to produce as proofs that Petrolia is no humbug, but a magnificent reality. And it may be added, neither is dealing in lottery policies to those who follow the avocation. Millions of money await his word of command. Thousands of fortune-hunters esteem it a favor to be by the great Fortunatus relieved of their greenbacks. To insinuate that *he* would take advantage of them by selling them mountain lands or pocketing fifty thousand dollars for his services—*faugh!*

An acquaintance of mine was once strolling through the heart of Petrolia, seeing what he could see. Stopping at a well, he inquired of a man who appeared to be proprietor its ordinary yield, and was told twenty barrels a day. The owner having got the impression that the other was a correspondent, addressed him thus: "I want to sell her; and if you'll give her such a notice in the paper as will bring me a customer at ——— dollars, I will give you —— per cent for your share. . . You won't? Well, what *will* you take?" If he had not understood the profession to be venal, would he have made the proposition so barefacedly, knowing that its actual yield was less than one-half of the quantity stated?

In *rating* the productiveness of wells there is apt to be so much exaggeration that the initiated invariably reduce the figures given, in answer to such inquiries, one-third or one-half, sometimes even two-thirds. In falsifying, however, some regard is paid to the semblance of truth, in order probably to avoid legal proceedings afterward. On this account, what wears the mask of veracity is more dangerous than glaring falsehood. Thus, the term "yield," in speaking of a well, may comprise brine as well as oil, the proportions of each to each ranging from one to ninety-nine per cent. Now, it is exceedingly difficult for a new-comer to estimate accurately the ratio of water to pure petroleum, and the matter is likely enough to escape his notice altogether. Again, it happens sometimes that a well is suffered to remain idle a day or a week, at the end of which it is pumped out, and "rated" at the product for *that* day. To an inquiry as to what it is doing, the stranger is carelessly told: "Yesterday, when we tested her, she gave forty-five barrels." Its average product may not be ten. A less transparent mode of swindling is not to test its actual yield at all, *when on the decline*, but to rate it at what it was in its palmiest days, using the conveniently loose phrase: "She's a two hundred barrel well." In the absence of a recent and accurate test, the person in search of information may certainly infer that a decline has set in, otherwise the telegraph would have kept talking about the Big Squirt, within the last forty-eight hours. The maxim that silence gives consent must be interpreted by the rule of contraries in the oil regions.

If I have thus exposed certain of the less notorious artifices by which the unsuspecting, and even some of the most wary, are fleeced, it is not with the design of warring on Petrolia or its master-spirits, much less retarding its prosperity in future. For it is indisputable that no well will sympathize with its owners' reputation merely to the extent of one pint in seven years. And the abiding prosperity of that section of country, with the true welfare of its population, will be promoted, not damaged, by putting a stop to the devices by which unsuspecting men and women, even widows and orphans, are stripped, and scalped, and flayed, and picked to the bone by a generation of sharpers. These of right no more constitute the people of Petrolia than did the slave-owning aristocracy of Dixie constitute "the South." Whether the acre of land shall sell for five hundred or five thousand dollars matters nothing to the aggregate production of oil; and ultimately this production is likely rather to increase than to fall off from an exposure of *how strangers are taken in.*

CHAPTER IX.

OUGHT I TO INVEST IN PETROLIA, AND HOW?

At last we have obtained, in the shape of facts and figures, something definite, tangible, trustworthy, as to the productiveness of the Pennsylvania oil region. In the month of April the wells actually yielded eight thousand eight hundred and fifty barrels per day, which is equivaent to three millions of barrels a year—to be increased at least half a million by additions during the spring months. I have no hesitation in asserting that the production for 1865 promises to be nearer four millions than three; and, indeed, may amount to all of the former figure. But without speculating too much as to the future, let us proceed to reckon up values upon facts as they are.

During the past five years the price of petroleum at the wells has ranged from ten cents to thirteen dollars and fifty cents per barrel. In the latter part of 1861, owing to the outbreak of a large number of flowing wells, bringing the supply suddenly up from one hundred and fifty to three thousand or four thousand barrels per day, the market value sunk so low that the cost of the barrel exceeded that of its contents; indeed, for illuminating oil, twenty-five cents per barrel was thought a high figure. This extraordinary cheapness had the effect of forcing the article into all parts of the country open to domestic trade, and even into Europe, where a large demand sprang up. In 1862 there was a considerable improvement in prices, but in the spring of 1863 they again receded from ten dollars to thirty or forty cents per barrel, in consequence of large numbers of flowing wells beginning to pour forth their liquid treasures. From that time an advance took place, and continued steadily till the middle of last winter, when it stopped at thirteen dollars and fifty cents. Various causes, the principal of which was a panic in the money-market, made prices again recede; and in April illuminating oil sold for only three dollars per barrel at the wells. Since then the movement has been slowly but gradually upward.

I shall assume, then, that during this period the *average* price has been midway between the extremes, namely, ten cents and thirteen dollars and fifty cents, which would be equivalent to six dollars and eighty cents* per barrel. The times in which it has been below that figure will not be found to differ materially in duration from those in which it has been above. As a rule, producers dispose of their oil from week to week, their tankage being insufficient to retain quantities for a long time on hand. Besides, the danger from fire and flood is such that if it were possible to hoard up, it would not be judged advisable.

Taking this as a standard, then, it will be seen that the

* Perhaps an extra allowance of one or two cents per barrel should be made for the higher price paid for *lubricating* oil on French Creek, and a small section of the Alleghany. But as the entire product of this quality has probably never gone up to one hundred and fifty barrels per day, and is now under one hundred, it was not thought worth while to take the difference into account, the values being only approximations to the actual figures.

actual yield in April was worth the sum of sixty thousand one hundred and eighty dollars per day, or twenty million four hundred and sixty-one thousand three hundred dollars per annum. Taking into account the net increase to be made during the spring months, the gross receipts of Petrolia, from oil only, during the year, would amount to twenty-three million eight hundred thousand dollars, and *may* run up, in 1865, to twenty-five millions.

So much for the crude article, delivered at the works; but it is manifest that the additional value given by refining the oil, and the cost of barrels, transportation, storage, and the like, enter into the regular, legitimate receipts of the Petrolians from this product. Ten millions of dollars would not be a too liberal estimate for all these services, making the yearly receipts from the oil business between thirty and thirty-five millions of dollars.

But in truth this has not been its principal source of income at some times. For the enormous profits realized by land speculators, during periods of excitement, such as prevailed last winter; the flood of strangers pouring into the valleys from every direction, with the mechanics and laborers seeking employment—all spending money lavishly, and most of it remaining in that country as profits —have added almost incalculable sums to the vast aggregate given above; and making the grand total, one year with another, probably not far from fifty millions of dollars per annum—a sum equal to the receipts of the general government twenty years ago.

Up to this time the most diverse as well as extravagant estimates of the productions of Petrolia have been set afloat by writers, who seemed to have adopted the maxim, in such matters: "Guess at half the amount and then multiply by two!" One of these "authorities" puts the production of 1865 at the modest sum of seventy-five million dollars, of which probably not more than two million dollars will be paid out as working expenses! It is patent that *he* has never been within one hundred miles of the country, or that he is doing his best to mislead the public. In accordance with these magnificent calculations, he sets down the capital invested in petroleum wells, lands, etc., at four hundred and fifty million dollars. It might just as correctly have been set down at a round billion, which would have been still more easily remembered. Every person knows that the nominal capital of most companies affords no index to the money actually paid in, making no allowance for the shares distributed among "friends at court," in order to secure their influence or their names as officers. If the profits, both gross and net, have been magnified out of all just proportion, there can be no doubt that the estimated amount of capital invested has gone through a like process. "The annual amount of the oil-product from the old wells," he goes on to say, "is sufficient to pay over twelve per cent per annum on the aggregate total of four hundred and fifty million dollars, estimated nominal capital invested in the business." I have no patience in reading such *stuff*, and shall not weary the reader's patience in attempting to refute it.

The Philadelphia Board of Trade, in their annual report for 1864, estimate the value of the product of petroleum for that year at forty-six million nine hundred and nineteen thousand four hundred and thirty dollars, or only sixteen per cent less than that of the iron product of the state, and two-thirds of the value of the coal-product. A decided drawback on these figures is that the learned statistician omitted *the odd cents!* The former writer guessed the product, and multiplied it by three; the latter stopped with doubling the figures!

Number Three steps forward, and gravely announces that all those estimates are huge exaggerations, the yield being now not more than six thousand barrels per day, against ten thousand or twelve thousand in 1862. This would make the annual value of petroleum at the wells amount to thirteen million seven hundred and ninety thousand four hundred dollars, or a little more than *one-sixth* of the first estimate! Even if we add fifty per cent for refining, transporting, etc., the total would be only twenty and a half millions, *minus* the operating expenses, government taxes, and the cost of replacement.

I have no means of estimating the productions of former years, and hence institute no comparisons as to increase or decrease. I am, however, fully satisfied that no such falling off (forty to fifty per cent) has taken place as is represented, and perhaps no falling off at all. Indeed, considering the veins opened on Cherry Run, in 1864, it would seem that there must have been rather a gain than a loss. The tables of exportation, at all events, do not warrant the belief that there has been a decrease. In 1862 the number of gallons of the oil-product, including crude, refined, and naphtha, was ten million three hundred and eighty-seven thousand seven hundred and one; in 1863, twenty-eight million two hundred and fifty thousand seven hundred and twenty-one; and in 1864, thirty-one million seven hundred and ninety-two thousand nine hundred and seventy-two. From the single port of New-York alone last year were exported five hundred and thirty-three thousand three hundred and ninety-four barrels, against four hundred and eighty-eight thousand six hundred and ninety the preceding year; and from all other ports, two hundred and sixty-one thousand three hundred and nineteen, against two hundred and seventeen thousand five hundred and seventy-six barrels in 1863. There is no likelihood that the exportations of 1865 will fall behind those of last year, but the reverse; while the home consumption will certainly be on the increase. With these data before us, and the knowledge that a heavy fall in the price of petroleum has taken place this season, we may dismiss all fears about a diminished yield having already begun.

The statements given above show what a degree of looseness has hitherto prevailed in making calculations of incomes, profits, business prospects, and the like. In fact, every thing has been conducted in the most reckless, hap-hazard manner imaginable, as if it were as difficult to arrive at the actual statistics of production as it is to strike an oil-vein from superficial examinations only. The author of *Derrick and Drill*—a volume which, whatever the slovenliness of its arrangement, was manifestly not written with a view to deceive—assumes that there are "no reliable statistics," and gives up the attempt to discover the truth in despair.

While I have estimated the actual receipts of Petrolia, from the oil business only, at between thirty million and thirty-five million dollars a year, and those arising from the sale of crude oil at the wells, at twenty million five hundred thousand dollars, with a probable increase to nearly twenty-four million dollars, it will not do to reckon this as such a rate per cent of dividend on so much capital invested. For, first, there must be deducted therefrom *working ex-*

penses; and, second, an amount sufficing to constitute a *renewal fund*—that is, to replace the old well when it gives out, by a new one equally productive. Thirdly, there is the government tax of one dollar per barrel. These matters have never been reduced to any thing like a system; and I am well aware that, in attempting to discuss them here, I am rather opening questions than settling them. To arrive at accuracy in the matter, twelve months' examination of the country and the history of its wells would be requisite.

Of the whole number of works on my list, (three hundred and twelve) only fifty-two are marked as flowing wells. I have not been sufficiently careful in every case to mark the distinction on my memorandum-books, but am positive that the total of these "institutions" is under seventy, and believe it to be less than sixty.

Let us assume, then, the last as the actual number. This would leave, at the time of my visit, the number of pumping wells yielding oil at about two hundred and fifty. But to this should be added those undergoing repairs and those exhausting the water—at least one hundred in all, and making three hundred and fifty requiring fuel, repairs, and wages for attendance.

Here again, however, we confront "the law of lawlessness," a few of those wells running night and day, some seven days in the week, others using their own gas as fuel, several engines pumping two wells, while others perform their work in three or four hours per day. Taking these and other irregularities into account, it will be seen that, to arrive at an accurate estimate of average working expenses would involve no small amount of time and labor, to say nothing of the outlay. But an approximative estimate to each well of seven dollars per day for fuel, and fourteen dollars for superintendence, wages, and repairs, or one hundred and twenty-five dollars per week for each paying well, will not be found very wide of the mark. This sum, multiplied by three hundred and fifty, the number of pumping concerns, will give as the total working expenses—two million two hundred and seventy-five thousand dollars, to which should be added (say) two hundred and twenty-five thousand dollars for superintendence, wages, and repairs to the flowing wells, and making the aggregate outlays for operating expenses, *two millions and a half of dollars.*

The next item to be taken into account is wear and tear or the replacing of exhausted wells, as they give out, by new ones. Here again a number of circumstances must be taken into account, not likely to be noticed by the uninitiated. In the first place, numbers of wells in operation are not paying their way, but kept going either with the expectation of finding purchasers, or that they may turn out better in time to come. Thus, the one which has been three years in operation, and is now producing only as many barrels of common oil daily, even if pumped every day from January to December, is unprofitable, and might better have been abandoned long ago. It will not do to put down three years as the period of its productiveness, but only that in which it actually earned a profit above working expenses.

The prevalent idea that wells can be "rejuvenated" or "resuscitated," so as to be nearly as good as new, is essentially a mistaken one. The Sherman, at one time flowing fifteen hundred barrels per day, has been resuscitated, by the blower, to the grade of fifty or sixty barrels. The blower and the pump have both failed to rejuvenate the Noble. The great Philips is no longer Philips drunk, but Philips sober, pumping fifty barrels instead of flowing nearly four thousand. How long even this may continue, is simply matter of conjecture. The best of wells, after deepening, re-reaming, and using every known appliance to force up the oil, have failed to become even second-class concerns. Their sources of supply are manifestly rather the leakings and oozings out of thousands of little crevices in the subterranean world than the fountains which leaped upward so impetuously when first tapped.

From the best information accessible, I am led to estimate the average period of a good well's productiveness at eighteen months. In a majority of instances the repairing and tinkering done after that date does not pay. In any case the purchaser should be careful not to accept the first day or two's yield thereafter as evidence of what it will be on the average. Now, to reach *one* productive source, with all the knowledge that has been gathered, it is necessary to put down *five* wells. Some persons say that one-fourth of those now sunk become profitable; but taking the entire oil region, I am within the mark in setting down the ratio at one in five. Now, assuming that machinery has only to be provided for the first, and that the depth will be five hundred feet, the cost of renewals will amount to twenty-five thousand dollars. This, multiplied by four hundred, the number of works already or prospectively yielding, would require an outlay, every eighteen months, of ten million dollars. I do not see how this heavy drawback can be safely reduced below *five millions of dollars* per annum, which would imply renewal only once in two years. In point of fact, the newly invested capital in the oil region is used to replace the old works as they give out, much of the original capital having disappeared for ever, either as unearned dividends, or gone to the wrong side of the profit and loss account.

Lastly, we have the government tax of one dollar per barrel on crude petroleum. There is a short and easy method of arriving at the amount of that: If the yield for 1865 be three million five hundred thousand barrels, the excise duty will amount to precisely *three millions and a half of dollars.*

We are now in a position to cast up the account, as follows:

Value of the crude oil at the wells, (say,)...........		$24,000,000
Operating expenses,....................	$2,500,000	
Cost of replacing works,................	5,000,000	
Government excise,......................	3,500,000	11,000,000
Net profit, (say,)..........................		$13,000,000

This would pay seven per cent per annum, on a *bona-fide* capital of one hundred and eighty-five million seven hundred and fourteen thousand dollars.

The political and commercial aspects of the subject remain to be considered, and to these I propose to devote a few words only.

Had the demand for American breadstuffs in Europe continued, during 1863 and 1864, as active as during the two previous years, it appears to me very questionable whether the diversion of so much capital and labor from the fields to the oil regions would have proved beneficial to us as a nation, in keeping our exports at about an equilibrium with our imports. It is not probable that the receipts, in currency or its equivalent, amounted to twenty-five million dollars last year, for crude and refined oil, naphtha, or other products of petroleum. As a people, we would have been quite as much enriched by twenty-five millions or its equivalent in return for wheat, flour, or corn, cotton, tobacco, or rice, as for petroleum, and no more. But it happened, very fortunately for us, as respects our trade with foreign countries, that about the time when the demand for breadstuffs fell off, petroleum found its way into market. By exporting it in such quantities, our government was enabled to make extensive purchases of war material abroad, without exporting the precious metals in such quantity as might have generated a panic at home, and thus have brought its own credit into disrepute at the moment when the scale began to turn decisively against the rebellion. As an instrument for buoying up the hopes of the people during a most terrible crisis, I view its effects as much more advantageous than in the mere money value of the exported article. For it is a serious matter to withdraw thousands of workers from the field and the shop, at a time when all products are selling for two or three prices—partly, it is true, through currency derangements, but partly also from the fact of such a diminished supply of the necessaries of life offering, that speculators are enabled to buy and hoard them up, compelling the government to pay largely increased prices, and thus heap up the aggregate of national indebtedness.

The total exports of this article in its several conditions, during the last three years, were as follows:

YEAR.	GALLONS.
1862,	10,387,701
1863,	28,250,721
1864,	31,792,972
Total,	70,431,394

The quantities of crude oil and naphtha were comparatively small. If we average the price last year at sixty cents, the product will be nineteen million seventy-five thousand seven hundred and eighty-three dollars. If we take the whole seventy million four hundred and thirty-one thousand three hundred and ninety-four at the same rate, the product will be forty-two million two hundred and fifty-eight thousand eight hundred and thirty-eight dollars —a respectable amount certainly, yet a moderate percentage on our entire exports during the past three years.

The amount of government revenue on crude oil for the year ending March 31, 1866, I have estimated at three million five hundred thousand dollars, on the assumptions, first, that the yield of the new wells will equal the decline on old ones; and, second, that the correct figures will be returned in to the government agents. That at many works the owners or managers will resort to every species of device to avoid payment of the heavy and almost oppressive impost, is what every person intimately acquainted with life and manners there will naturally expect. If, however, the officers experience any trouble, or have reason to believe that misrepresentations have been made, a very simple and efficient remedy offers itself. PUBLISH THE FIGURES OF EACH WELL EVERY MONTH! It would be a terrible infliction, but one richly deserved. While essentially right and fair in itself, the measure would purify the stifling atmosphere of Petrolia, and for that matter, that of the whole country. Weak-kneed interests would growl over it more than the imposition of one dollar per barrel of tax; but the people at large would be greatly benefited through the instruction thus received.

That the wells in process of sinking will yield enough

to make the supply keep up with what it now is, I have no manner of doubt. The prospect for 1865 is, that they will do considerably more than this; for 1866, that they will at least equal it; beyond that year it is not advisable to venture on calculations or estimates of production in a field where the law of lawlessness has all along prevailed.

If we add to the revenue to be derived on crude oil, that which is already being received upon the refined article, (twenty cents per gallon,) the aggregate receipts into the United States Treasury cannot fall below twenty million dollars, and may amount to twenty-five millions, during the twelve months stated. Of course, all this comes out of the pockets of the American people as consumers; yet there is perhaps no other source, yielding so much revenue, where the load would be felt less oppressive than on this article, as is manifest from the fact that petroleum not only maintains its sway, in spite of these heavy impositions, but is extending it in every direction.

It may be observed here that, by the present law, the greater portion of the lubricating oil nearly escapes taxation, the excise duty of one dollar per barrel on the crude article being barely three per cent on the selling value at the wells. As most of it is used in that condition, it is manifest that the revenue suffers a considerable loss, or else the imposition is too heavy on the illuminating kind. A thorough revision of the whole subject is one of the first matters to which Congress should give its attention.

I have been asked, scores of times, whether petroleum is likely to be found elsewhere, particularly on the great slope east of the Alleghanies. Where the most distinguished *savans* have been found at fault, the discovery marking a new era in geological science, the mere learner may well be excused from taking his finger from his lips until experiments have settled the question. I shall, however, venture a few *opinions*, which may be taken for what they are worth. In the first place, then, those great divisions of the secondary formation, known as the Cambrian, the Devonian, the Silurian, and the carboniferous, are not known to exist east of the Blue Ridge from Virginia to New-York. The Silurian and the Devonian abound in North-western New-Jersey and that portion of Pennsylvania on the opposite side of the Delaware. Second, the groups known as the Chemung and Portage rocks, (the former composed of highly fossiliferous shales with thin-bedded sandstones, and the latter of flag-stones and shales,) which abound in the oil regions *above* the river-beds, send only a small arm into the eastern part of New-York, touching the Hudson near Catskill. With that exception, there exists none of these groups east of the head-waters of the Delaware. Third, sufficient information has not yet been obtained respecting the *inferior sand-rocks* of Venango county in which oil is obtained, whether to identify them with the Onondaga, the Helderberg, the Hamilton, the Clinton or the Medina group, all coming to the surface in Western New-York; but it seems probable that the oil-bearing rock extends a considerable distance in that direction, and may enter the Empire State from the south-west. A much better guide, however, will be to get *large fragments* of the second, third, or fourth sand-rocks from below, and compare these with rocks near the surface in New-York, than to trust to mere surface resemblances or dislocations, which are of no account whatever. Fourth, we know that the sand-rocks of West-

Virginia, as also those of Eastern Kentucky and Eastern Ohio, yield petroleum at the average elevation of fifteen hundred or two thousand feet, geologically, above the oil-bearing measures of Venango county. How far beyond Kentucky or how far west of Ohio this liquid may be discovered, time alone will show; but it seems likely enough to be obtained in parts of Tennessee, Alabama, and a very considerable portion of the North-west. Fifth, it is certain that the sandstones and shales of New-Jersey and the sandstones of Connecticut have no sort of affinity with those of Western Pennsylvania, where the shales are principally of clay and full of fossils, and the sandstones or conglomerates are more or less fossiliferous and arranged in thin layers, with hardly a perceptible inclination. Sixth, *I think* it will be found that the true source of petroleum is not the sandstone or arenaceous limestone, but the shales which are in places bituminous and in places petroleous, (to coin a word.) By what particular process the carbon, originally mixed with the clay-beds, (now shales,) was distilled into oil and gas, and then injected into the crevices of the harder rocks above, I do not pretend to know; and possibly it will never be known until deep shafts have been sunk to the depth of one thousand or fifteen hundred feet.

But of one thing we may rest assured, namely, that there is quite as good a chance of striking a rich deposit of Orange county milk, or even butter and cheese, at any point east of the Blue Ridge in Virginia, Maryland, Pennsylvania, New-Jersey, and New-York, as in opening a ten-barrel vein of petroleum. There is abundance of wealth in the mountains and valleys of that section, besides rock-oil; but as the old-fashioned article does not possess the charm of novelty, nor even that of distance, to "lend enchantment to the view," it is despised. The millions of money which, if judiciously invested there, might double themselves in a few years, are transported to Western Pennsylvania and West-Virginia, to add to the intoxication already too prevalent, and come back—if in hundreds of thousands, the owners will have good cause to bless their stars.

Men's theories, as openly avowed, are apt to be modified by their interests, and their calculations as to the future by their theories. Hence it is found that the naturally sanguine and the mercenary advocate the notion that, in the great laboratories of Nature, chemical and mechanical agencies are at work, by which petroleum is constantly generated, and hence that there need be no fear of exhaustion. They point to Babylonia, Birmah, and Trinidad as proofs; but it is manifest that Petrolia does not adopt the fashions set abroad; and indeed does not follow any set fashion, either at home or abroad. It recognizes no rule, no precedent; but follows its own strong, wild impulses from day to day. *That* argument, therefore, will be found very childish against the hard fact that hundreds and thousands of wells have already dried up, and cannot be made to reproduce *oil;* though salt brine flows profusely enough. Attribute this falling off to the lack of gas or of pressure from above, as we may, the painful truth forces itself upon our understandings, that petroleum *is not there*, and cannot be coaxed, wheedled, or forced to the surface, for the best of reasons.

On the other hand, I think, they err who hold that because individual wells have ceased to yield, therefore the production of petroleum is likely to stop, one of these days, as suddenly as it began. Facts show that the area of oil-bearing territory is not confined to the "pent-up Utica" of any creek or river bottom. A careful examination of the country shows, further, that the bottoms, *as such*, have no connection whatever with the deposition of oil; though they have had some with its discovery. But, making allowance for the greater cost of boring in the first place, and of operating expenses, in the second, I have no doubt that the hills will yield as freely as the low-lands. So believing, I have so taught. My mind has neither been influenced by the fear of a sudden cessation, nor the belief of a constant creation of oil to any great extent. Petrolia has not appeared to me in rainbow-hues, as viewed from a distance; nor a gray, drizzly mass, as viewed from within. With land and labor at reasonable rates, it is likely enough to be good for seven per cent per annum—perhaps even ten; though if I were a stranger, I should decidedly prefer a United States seven-thirty bond.

"Then you maintain," say some readers, "that it is foolish to invest a dollar in oil stocks, to sink a well, or to organize a company with a view to the further development of that region. For if the business be in many cases a swindle, a system of deception and falsehood, as you represent, and but a poor affair at the best, no prudent man will have any thing to do with it, except to get out of it as quickly as possible."

As at present carried on by many, I have no hesitation in pronouncing the enterprise a gigantic system of wrong; but surely this is no reason why good and bad interests, honorable and dishonorable men, should suffer together. Nay, it should be the desire of every individual and every company conducting business in a straightforward manner, to have the villainies of their unprincipled neighbors exposed. The sooner this is done the better for themselves and the community. For it is as certain as sunlight that roguery will be detected *some* day; and the longer that is put off, its consequences will reach the further, so many more persons will be involved in the guilt, so much less rigid will be public sentiment as to right and wrong. The reäction which set in with this spring freshet might perhaps have been put off by some newly organized method of deception; but with a shoal of fresh fortune-hunters sent into the oil region, mostly to return *minus* the contents of their pocket-books and their balances at bank, a fresh and deeper outburst of indignation would have arisen against Petrolia and all connected with it. It is for the salvation of what is genuine in that country that I have made some of these exposures. It may disarrange the calculations of some, who have expected that "to-morrow would be as this day, but more abundant;" but it is one of the punishments of injustice that the pillars upholding it are liable to be knocked away, and the edifice to tumble into ruin at any moment—always the *wrong* moment, of course. Samson is not to be blamed for tugging at the pillars of the temple; but let the gaping and jubilant Philistines on its roof look out for their personal safety.

Every enterprise that proved pecuniarily profitable at first has been almost ridden to death by knaves or fools for a time. Look at the history of the railroad interest, both in Great Britain and this country; at telegraphs, commerce, manufactures, even agriculture. In all a few sagacious individuals have originally done well, on learning which excitements have sprung up, and for a season it looked as if all creation were rushing in that direction. The American mind is peculiarly subject to these financial whirlwinds. Ready enough to suspect sordid motives in others, it puts the same implicit faith in its frothy newspaper articles or letters, as if it never had been misguided by those so-called leaders of public opinion. He is not the people's friend who fails to point out this liability to go wrong as well as right, by rushing in droves toward some common object.

As a rule, the judicious will suspect the existence of danger in whatever direction the crowd is driving. If in politics, it is likely to be impelled too far and too violently by leaders, who are often selfish, or who have let passion get the better of their judgment. If in any particular department of industry, it is likely so to glut the market that prices will fall to ruinous figures, leaving bankruptcy to restore the equilibrium between supply and demand. On two occasions only will the most sagacious financial boatman dip in his oar—when an enterprise is in its incipiency, and he foresees with certainty that it will be profitable; or after a general collapse, when fragments of the wreck can be gathered cheaply, and wisdom can be learned from the misfortunes of others. The intermediate stage, when the public mind is at fever-heat with high expectations, is one that ought to be selected for selling out and retiring from active participation. Of course, every man cannot act thus; but if a number of sensible men should do so, the step would have a most salutary effect in checking the blind rush of the thoughtless multitude.

Now, on three different occasions Petrolia has passed

through this process—before the great downfall in the price of oil, in the autumn of 1861, in the spring of 1863, and that of 1865. Previous to each of these lapses numerous operators had made fortunes, which turned the brain of the community. A strife sprang up as to who could first "stake a claim" in this new El Dorado, which as far surpassed California as the Golden State did the land of steady habits. The market was, presently or prospectively, glutted, and down tumbled prices—at two of these epochs to a ruinous degree. Multitudes of wells were thrown up in disgust, and a general exodus of population took place. This refluent wave has not been so great the present year, and will not be so great hereafter, because the demand for the article has immensely increased, and the business of boring for it has become one of the permanent interests of the country. At the same time, in proportion to its permanence, and the decrease of great fluctuations in price, its profitableness as a whole is reduced to a common level with all other legitimate enterprises. There is no escape from this social law, either in or out of Petrolia. Great fluctuations—great profits—great losses—great risks—and, it may be added, great immorality. *Per contra.* Increased demand—less risk—less profit or loss—less immorality in conducting it.

I may be told that this is altogether too sombre a view of an enterprise which has enriched and is now enriching thousands; which turned the rate of exchange in our favor, while a most expensive war was being waged; which promises to be like the widow's cruse, when the prophet directed her to go and pay what she owed and live, she and her son, with the residue; which came providentially, if not miraculously, to aid us in throttling the great rebellion. For the good that it has accomplished, I feel grateful. In the long run, one well the less will not be put down from an exposure of the frauds perpetrated by those concerned in it, nor will a gallon the less be exported. The country as such has nothing to fear from the truth, whatever those engaged in the work of deception may have. As to the enriching of individuals, a very pertinent inquiry arises—How much of that wealth is due to legitimate enterprise in raising petroleum, and how much in speculating in land, or taking advantage in some way of others' necessities? It has always been the boast of wrong to get its dividends first!

"But there are large companies," I may be told, "which have paid as dividends two, five, nay, ten times the amount of capital stock paid in. What will you say to *them?*" And here occurs an account of the Columbia Oil Company of Pittsburgh, which will serve as an illustration of perhaps a dozen others. The facts are taken from the *Trade Circular:* "The Columbia Oil Company was originally a firm, under the style of Ritchie, Hardie & Co., consisting of seven persons, who purchased the Story farm, containing nearly five hundred acres, at the outburst of the petroleum excitement in Venango county, for a few thousand dollars. In 1862 the company was organized, and purchased the Story farm from the firm of Ritchie, Hardie & Company for one hundred and twenty-eight thousand dollars. The company was organized with a capital of two hundred thousand dollars, divided into ten thousand shares of the par value of twenty dollars each. These shares sold in the market, during 1862, for from two to ten dollars. The shares gradually increased in value until, in March, 1864, they were worth one hundred and twenty-five dollars each. From the time of the organization in 1862, until March, 1864, dividends had been paid on the stock to the amount of three hundred thousand dollars. In April, 1864, a further dividend of eighty thousand dollars; in May, of one hundred thousand dollars; and in June, of one hundred thousand dollars were paid, or nearly three times the original value of the stock. In June, 1864, the stock was enlarged—that is, the original shares were called in, and fresh stock at the par value of fifty dollars per share, in the ratio of five shares of new stock for each original share, issued to the holder thereof. From June until August, the new stock advanced in price until, in the latter month, it was worth one hundred dollars per share, or double its par value; and the stockholders had received dividends, in July and August, to the amount of two hundred and twenty-five thousand dollars. The dividends for September, October, November, and December amounted to four hundred thousand dollars, being to parties who have held their original stock a payment of one hundred and twenty dollars per share in dividends in eighteen months. The person who paid, one year and a half ago, the original value of twenty dollars [per share] for one hundred shares, and has held his stock, has received twelve thousand dollars dividend up to December, and from the profits of increase of capital made in June last obtained an accession to his stock of four hundred shares, which shares, although of fifty dollars par, are now worth, with his original shares, forty-two thousand five hundred dollars, making a clear profit of fifty-four thousand five hundred dollars in eighteen months."

That is to say, the person who paid in two thousand dollars as stock has received over twelve thousand dollars as dividends, besides a multiplication of his shares by five, the new issue being one hundred per cent above par last December; but since then there has been a large fall in stocks, so that Columbia, which had reached one hundred dollars, as stated, has fallen to seventy dollars. This diminishes the "pile" of profits very materially.

Again, the Columbia Company purchased a farm which happened to be in the centre of the valley—the very core of the oil region. Three miles above and as many below it would not have enjoyed such a "streak of good luck."

But, thirdly, it is quite true that not a few firms and companies that went largely into the movement—purchasing at low prices large tracts of land, and leasing it out in small lots to others for one-half of the oil, with perhaps a large bonus besides—have succeeded to a degree almost unparalleled in these times of novel developments. What individual or association would not become princely rich, if owning in fee simple a few hundred acres of the Wash. McClintock, the Hyde and Egbert, or the Story farm? But the matter wears a different aspect when it is proposed to go to that country and lease property on the terms usually granted by the present owners, who are not likely to act from the principles of pure charity, as they are probably quite as familiar with the productiveness of their lands as strangers could be the next hour after their arrival!

In a word, it was one thing to organize a company, purchase lands, and bore for oil in 1862, and another in 1865.

Whether it is or is not advisable for a prudent man to embark in petroleum enterprises at present, is a point on which I refrain from giving an opinion. That many of both the new and the older wells will be amply remunerative, none but prejudice will attempt to deny. That a large amount of valuable experience has been gathered from the past is known to every body; and the effect of this, in the judgment of careful and conscientious operators, is to reduce the proportion of non-paying works from nine out of every ten to four out of five or three out of four. This consideration, however, is offset by another, namely, that no such productive wells have been opened the past two years as were during 1861 and the two following years. To tap a two thousand barrel spring was at one time a matter of not *very* unfrequent occurrence; there is not a well which yields over five hundred barrels to-day, only one gives over three hundred, and ten over one hundred. A two hundred barrel well is viewed with as much interest now as one flowing two thousand was three years ago.

Hence, while the ratio of paying works opened is on the increase, those mammoth concerns which poured out their thousands of barrels daily have disappeared, probably for ever. The risk is less, but so are the prizes. The enterprise is rising from the plane of a lottery toward that of certainty, and I may add, toward legitimacy and morality. So much the better for all concerned, not even excepting the land and stock gamblers.

To ask such indefinite questions as, "What do you think of the oil regions?" "What is your opinion of investing in petroleum stocks?" is childish in the extreme, it being impossible to give a *general* answer on the subject. In another chapter I have pointed out to what an extent "the law of lawlessness" prevails; and that principle holds good as to the productiveness of farms and the financial condition of companies. That one property is "dry" affords no criterion to the lot adjoining it, which may open with a fine flowing well; that one company is rotten to the core, being managed by rogues and sustained by dupes, affords no proof that another may not be sound and remunerative. One may pay out its working capital in dividends; another may scatter and yet increase. It is true that Tray suffered from being in Snap's company; and that the worthily managed have to bear a part of the odium which is attached to the worthless, when both are engaged in the same calling. The prudent stockholder will be as careful about sacrificing his shares during a time of panic or reverse, when errors in the management have been discovered, as he would be about investing in company with a crowd. In truth, his interest is more valuable after each instance of mismanagement or wrong-doing has been brought to light, and that is the time, above all others, when he should hold on to his interest. Let him address himself to manning the pumps, lightening the hold, and seizing the helm, not to throwing himself into the yawl at the first note of alarm.

But when I observe, at this late day, men who have been all their lives engaged in other pursuits crowding into the oil regions, and expecting to make fortunes speedily, as land speculators, agents, and even superintendents, competing with those who have served years of apprenticeship in these capacities; when I see those at home expending their means lavishly to sustain others equally inexperienced with themselves, I inwardly say: "Friend, you will have to pay dearly for your whistle before all is over; but probably no other lesson would have taught you needed wisdom, so profound was your belief in your own invincibility, after becoming well-to-do through measuring tape or paring cheese. Go forward and be taught!"

Under ordinary circumstances, we should hardly think of employing a butcher to do our preaching, or a farmer to purchase dry-goods at wholesale. Most persons have come to believe that a certain amount of training, of experience, as well as natural capacity, is requisite to employment in any situation requiring knowledge, skill, and judgment. Not so with many of the oil companies. An officer who has done his duty bravely on the field, a glib-tongued lawyer or politician, a broken-down merchant—any body, in fact, who is "smart" enough to button-hole a board of ignorant directors—may be safely intrusted with the charge of a heavy interest in Petrolia. I am not impugning the characters or capabilities of superintendents as a class; yet those who have been longest there will most readily concede that at first they were so verdant as to be unfit for the trust reposed in them. When the enterprise was in its infancy, the readiest and likeliest had to be selected as managers. But surely the history of six years has not been without its fruits in training a body of practical men, whose experience of the various operations and processes to be gone through is worth something. Yet scarcely a day elapses without witnessing utterly green hands coming out to take charge of important works—men who are, nevertheless, too indolent or self-opinionated to spend their leisure hours in gaining instruction from their more experienced neighbors.

And so, before forming an opinion as to the merits of any particular interest, it becomes necessary to examine into the character of the agent sent out to conduct its affairs—his soundness of judgment, his knowledge of the business, his integrity, his application. Does he spend most of his time on the ground, or is he to be heard blustering in the bar-rooms, or, it may be, using his own position to acquire wealth by dishonorable methods? Recently, in Titusville, a bureau of statistical and other information was started. I suggest that its agents make particular inquiry as to the calibre and conduct of those who are intrusted with the management of important works there! To lead a company triumphantly through the storm of battle is glorious; but it does not insure success in purchasing real estate and machinery, employing labor, locating wells, making contracts with persons who are up to every conceivable dodge and evasion. It is fine to display eminent ability in a law court, or to glisten on the editorial page; but it by no means follows that the genius will be crowned with laurel as an oil operator.

To the individual, therefore, desirous of investing there, the first consideration should be to procure full and accurate information as to the facts ascertained and the persons employed by companies. This he can do either by a personal inspection, which must not be too hurried, or by consulting those on whose word he can place implicit reliance. There are some such in the oil regions. I have found an agent who corrected too high figures which had been given me as to the yield of a well, against the interest of the company and himself as a stockholder. Nor are such men in minorities as lean as may be supposed. It takes time to separate the wheat from the chaff—to ascertain whose word is deserving of trust and whose is not.

A refiner, of several years' experience in that country, and in easy circumstances, remarked to me that he and his partner always invested their surplus profits in new wells, though they did not believe that more than one in ten returned first cost and operating expenses. Nay, he would go further, and as a business man would guarantee to pay fifty per cent per annum as dividends on moneys intrusted to him for investment. This is certainly a tempting return, and I have no doubt was uttered in good faith; but he did not state *for how many years* the obligation would hold good; and this is the point of all points in oil investments to be considered. He added that he would expend the money in localities selected by himself, and under the direction of men in whom he could confide.

Another observation made by the same person was this: Let the man who has only a moderate sum to spare, *distribute it among several interests*, instead of risking all in one, where he is liable to lose all. I can imagine no better piece of advice than this, the only obstacle to outsiders in carrying it out being the lack of information as to the character of companies having shares for sale. But such an inquiry must be made about *all* incorporated concerns coming before the public for means, or woe betide the unlucky wight who trusts to chance. The man who invests in any enterprise blindly, does what in him lies to demoralize business, by holding out temptations to defraud. The duty, on moral grounds, of exercising strict vigilance in such matters, has not been sufficiently impressed upon the public by their professional teachers.

Among the methods by which, it seems to me, boring for oil might be properly and profitably conducted, is the following: From four to six persons might associate as a firm or a company, putting in a total capital of twelve thousand to fifteen thousand dollars. These men should consist of an experienced driller, an engineer, a carpenter, a tool-dresser, with perhaps a teamster, and a general manager, all ready to put down their hands and take vigorous hold at work. They select their acre-lot, which they should own in fee simple; erect their own derrick, engine-house, and machinery; haul their own lumber, fuel, and supplies; purchase their own engine and apparatus, then proceed to drill out their own well. Thus associated, they become their own grocers, boarding-house keepers, teamsters, clerks, etc., each having charge of his several department. If thought advisable, two relays could relieve each other, taking turns by day and night at the machinery. With a reasonable surplus to begin with, if one well proved unsuccessful, they could proceed with a second, a third, or even a tenth, until they "struck oil," if it should be strikeable. This done, a portion of the proceeds might be set apart to extending operations.

There are too many and too obvious advantages arising from such associative effort to need further elaboration here; and it only requires *one* attempt of the kind to be made in Petrolia for such a movement to become quite general. When this has been accomplished, companies, with their stocks "watered" to insipid weakness, with their costly superintendence at the works and management in the large cities, will find competitors in the producing field who can both undersell them and make a profit by so doing; at the same time the workmen, with this new avenue to preferment thrown open, will have a stimulus to steadiness, economy, and the acquisition of knowledge, such as they cannot have at present.

The excise duty of one dollar per barrel on crude oil will fall with almost oppressive weight upon this interest, in view of the effects of the late freshet and the reäction which has arisen in the public mind. Of the dead-and-alive concerns, kept going in order to find purchasers, I have no doubt that very many will suspend operations altogether. Probably one well out of every ten now pumping will thus get snuffed out before midsummer, after flickering a few weeks longer in the socket. The falling off thereby in the production of petroleum will not, however, exceed one or two per cent, except indirectly; but in that way its effects cannot fail to be serious, the imposition being from ten to thirty per cent on the gross product of an interest, which is certainly not netting ten per cent on the capital invested. Yet it is hard to sympathize with Petrolia in its misfortune, considering the systematic falsifying and swindling at which it has connived, if it did not directly encourage. Assuredly, if its inhabitants had been careful to represent matters precisely as they were, such a tax, in addition to twenty cents per gallon on the refined article, would not have been thought of. But the oil men gave encouragement to the belief that they could stand that or any other burthen, and so it was fastened on their shoulders. They are the sole authors of their own misfortunes, the architects of their own ruin, if ruinous the duty shall turn out to be. Often, on passing through the country, I could not fail to perceive the unpleasantness of their situation. Occasionally taken for a government official, I found the yield at which a well would be rated as much too low, as it was likely *not* to be when the true object of my visit was understood. Poor fellows! They are placed between the upper millstone of Uncle Sam and the lower of the public prints, between which their situation is not to be envied.

The annoyances connected with this impost are fully as

much a theme of complaint as the tax itself. It may be, however, that these are selected as a side-issue for attack, when the true object assailed is the one dollar per barrel, operators being fearful of confessing that it falls with crushing weight. Be this as it may, the duty must be paid by the owner or superintendent of the well within ten days, after rendering into the deputy assessor's office an account of its actual production, which must be done on the first, eleventh, and twenty-first days of every month. This necessitates at least three journeys in the month to that functionary's headquarters, and may require four or five, through a country not the best adapted in the world for pedestrian excursions, and where horse-flesh is an expensive luxury. It seems to me to be the obvious duty of the government to have its agents visit certain sub-districts on days specified, and *receive* the statements of production, if not the tax itself. With a load so exceedingly heavy it ought to offer every facility in its power for mitigating the effect; and it is to be hoped that, despite the folly which brought on such a visitation, Congress will see the necessity of lightening this burthen upon an interest which, *as a whole*, has doubtless sunk to the general level of American industry on the score of profitableness. There is no valid reason, when the truth becomes known, why Petrolia should be taxed from fifty to eighty per cent on its gross production, in addition to income and indirect taxes, while all other interests are rated, as nearly as may be, at from three to six.

I am not sure but the statement that "oil-mining" has been reduced to the general level of American industry, is a truism rather than a truth. In fact, the sea and the atmosphere do not more constantly find their own general levels than does business to ascend or descend to the general average. This may shock the notions of those who have been made dizzy by perusing the reports of certain great companies, or the productiveness of certain mammoth wells; but such must learn to take into account the thousands who have invested money and got back little or nothing, as well as the hundreds who have grown rich, and the tens who have become millionaires. They must bear in mind the disproportion of "dry territory" to that which bursts forth every little while with spouting wells. They must take in the disastrous years 1861, 1863, and 1865, equally with the prosperous ones which were sandwiched between these. And they must, if they would arrive at truth, make a note of the secret disgust, the silent disappointment endured, as truly as of the demonstrative displays made by those who have made their fortunes by a single throw. When the dark background as well as the bright coloring of this oil-painting has been contemplated, it will be seen that it differs from other pictures only in the *intensity* of contrast, not in the *proportion* of light to shade. The bright lines are more brilliant, as the sable portion is more pitchy dark than other works of the kind. With a region close by our doors, with free exit and entrance, with every facility for acquiring information about it, how could it, indeed, be otherwise? Common-sense and the principles of sound political economy teach, *à priori*, what I have demonstrated by figures.

CHAPTER X.

PRACTICAL CONSIDERATIONS.

SINCE the discovery of petroleum in Venango county, the state of Pennsylvania has done nothing to assist in its development, although deriving a large revenue therefrom. There is an urgent necessity for immediate reform in this regard. From the outset, for example, a state agent ought to have been appointed, with authority to collect statistics of the wells' productions, arrange, and publish them. That this should have been so long neglected, is alike discreditable to Pennsylvania and the interests more immediately concerned. At this time of the day, the propriety of gathering the statistics of all great departments of industry, fortunately requires no extended argumentation; but in a new growth of the sort, where the public mind was so liable to be led astray by falsehood, and where a dangerous reäction was sure to set in after the discovery of systematic misrepresentation, the case was so much the more urgent. Had monthly or even quarterly returns of the actual yield of those fountains, under oath, been made compulsory, the fever of excitement would not, it is true, have risen to so dangerous a pitch, but neither would it have been followed by a corresponding prostration. It is supposed that this oversight will be remedied by the general government requiring such returns thrice a month; but as the assessor is not obligated to publish the amount of taxes collected on crude petroleum from month to month, the public at large will have no access, within reasonable time, if at all, to full and trustworthy information through this avenue. Favored individuals or newspapers, it is true, may procure this on the sly; but partiality in such matters is exceedingly mischievous, as it is essentially unfair. The earnest attention of the state authorities is, therefore, asked to this matter.

Again, while Petrolia is inhabited by an essentially peaceable and orderly population, no protection exists to property-owners against careless or malicious persons, who withdraw the tubes and leave their wells open, to render unproductive perhaps half the works on a large farm. Along all the lower portion of Oil Creek, particularly Tarr farm, I have elsewhere pointed out the calamitous effect of this careless selfishness or a disposition still more objectionable. Beyond a question, the aggregate productiveness of the oil region is lessened one thousand barrels per day from this cause only. The matter is one coming properly under legislative cognizance, and should receive immediate attention. I more than suspect that hitherto the companies most interested in having such abuses corrected, have been afraid to move in the matter, lest the admission might expose "the nakedness of the land" before the public, chiefly at the great money centres. They have chosen to keep on suffering that they might have the privilege of uninterrupted sinning!

The small outlay connected with carrying these suggestions into effect could be borne by a tax on each productive well, or upon the capital invested by each company, firm, or individual.

Once more, it should be the duty of such a state agent to require the constructing and keeping in proper repair of suitable highways, bridges, sidewalks, etc. This is a matter of prime necessity. Extraordinary abuses call for extraordinary remedies. At Oil City, for example, the municipal authorities seem neither to know nor care for such matters; the attention of men of means is too much absorbed in the *race-course* to descend to such a pitiful level as a common road. Outside of the borough there is no law recognized. The short-sightedness of individuals and companies has literally converted the Petrolians into a community of murderers; for I can regard the systematic, persistent cruelty and oppression shown to the horse, as at least "murder in the second degree." Teamsters assign six months as the average period for using up the finest quadruped that ever toiled on man's behalf. This harsh treatment brutalizes drivers, as it benumbs the best feelings of every looker-on. The compelling of all landowners to lay out proper roads and keep them in suitable condition, for which abundant materials are everywhere close at hand, would prove a vast saving pecuniarily to all concerned at the year's end; but until all are forced to act, it is idle to expect any one person or interest to move in the matter.

Legal enactments are also needed to provide for the construction of levees, drains, etc., as well as roads. It is true, government machinery is usually heavy and complicated, as compared with individual action; but where, as in the region under consideration, individuals and associations refuse to move a step in the prosecution of improvements, which would enure to the benefit of all, it becomes the duty of the law-making power to interfere, perform the work in the most expeditious and efficient way it can, and charge its cost upon the interests thus benefited, or rather rescued from destruction.

It is little to the credit of the "Keystone State" that no systematic and thorough examination of the oil region should have been undertaken on its behalf by scientific men. The *Geological Survey of Pennsylvania* is a noble work, unfortunately published shortly before the extraction of petroleum. When an interest has sprung into existence, with taxable property annually increasing at the rate of many millions of dollars, some consideration is due it by the constituted authorities of a great and enterprising state like Pennsylvania. The best talent, joined to the utmost disinterestedness of purpose, ought long since to have been secured and kept in the oil region to collect every specimen, and record every observation made by practical men, assisting the latter, whenever necessary, or rectifying their mistakes in matters of fact or inferences. With all thus working harmoniously together, a body of scientific truths, of the utmost value in a practical point of view, might have been collected and published by this time, instead of the chaotic mass of facts, opinions, and endless contradictions, which have arisen out of Petrolia, as dark and dense as the smoke of the bottomless pit. Such an agency would have cost a few thousand dollars, which could easily have been assessed upon the interests benefited; while it would doubtless have added millions to the permanent wealth of the state, and thus indirectly have been a source of revenue. It is not yet too late to begin this most desirable undertaking; but not an hour should be lost in setting about it. So far all is "confusion worse confounded" in this respect. The accounts and theories of men are as diverse as the characteristics of the wells. The upper and nether worlds are for once in most delightful accord, and man lives according to nature. Scarcely two persons will be found to agree in their versions of either facts, phenomena, or causes. The rock which one styles a sandstone, another terms a limestone. What A denominates soapstone, B calls a tenacious clay. Theories are as dissimilar as men's faces, every body ridiculing that entertained by every body else. No wonder; for Petrolia does not contain twenty men who have had a good scientific training, and acquired accurate and diligent habits of observation. The state authorities must undertake the task of reducing this chaos to order, of reconstructing these *disjecta membra* into a system of utility and beauty, or it will never be done properly.

While the state is under obligation for duly protecting and extending this great interest, the public press of the country should bestow on it more earnest attention than has yet been given to it. A sufficient excuse for such oversight hitherto has been the frightful civil war, which involved most of our leading dailies in enormous expenses. No such excuse, however, can be justly offered on behalf of journals professedly established to throw light upon Petrolia. Every such paper should have one or more able and trustworthy correspondents in the field—men who would religiously keep their fingers from being soiled with a bribe. Unfortunately it has long been remarked that journals "devoted to" certain specialties have not been *devoted* at all; their conductors satisfying themselves with facts rehashed from the daily papers, with some common-place observations to fill up a column.

The petroleum journals have an opportunity of wiping away this reproach; and I learn with pleasure that some of them are striving to do so.

Allusion has already been made to the cost of keeping a pumping well in operation. Unless gas be freely given off, it requires nearly a cord of wood to supply the furnace for twelve hours, costing from seven to ten dollars. Coal is rather more costly, especially in the winter season. Add to the outlay for fuel the salary paid a superintendent, the wages of an engineer and a laborer, besides occasional outlays for repairs, and the total working expenses can hardly be set down at less than one hundred and twenty-five dollars per week, for a work of ordinary productiveness. This absorbs the receipts, at the average price of six dollars and eighty cents, of three and a half barrels of illuminating oil to begin with, assuming that the well is pumped steadily all the year round; otherwise, the proportion of loss is still larger. Seldom does a four or five barrel well, after paying the excise, do more than clear working expenses, in the absence of gas as fuel.

To reduce these the engineer is, in some instances, required to make himself a man of all work, occasionally dressing tools, repairing the sheds, or repairing the machinery which may have got out of order. The superintendent of a small concern also performs often the functions of clerk and laborer, in addition to his own, or he hires out a portion of his time to other companies or individuals. One man may thus have the general oversight of half a dozen or more contiguous interests. The work of barrelling the oil requires little labor and (commonly) too little time. Refiners or their agents ordinarily go from farm to farm and purchase it on the ground, sending their own boats or teams with barrels for its reception. In this case it sometimes needs a degree of watchfulness, as the purchaser may have his barrel ever so little above the size, all the excess over forty gallons being so much clear gain to him. On this account, some prefer to sell the petroleum in bulk only. In all these respects I have no doubt much saving may be effected, and, further, that the time for turning a new leaf on the score of economy has fully arrived.

An intelligent operator lays down the following rules for the guidance of superintendents and other employés: *Be careful in every thing.* Measure, calculate, estimate, weigh accurately in advance; and when action is required, let nothing be done rashly. A well requires to be as tenderly dealt with as a nurse's patient. *Let well enough alone.* If your well be yielding ten or twenty barrels per day steadily, don't undertake to make new and rash experiments upon it, even though they may have succeeded elsewhere. Never count upon two wells being quite alike in this respect. What is sauce for the goose in Petrolia may *not* be sauce for the gander. *Look carefully to the pump-valves.* The sudden stoppage in the yield of productive wells is more frequently due to disorder in the valves than to any other cause.

Of mercantile business in Petrolia the same remark is usually made as as about boring for oil: The man who is adapted to it and conducts it on correct principles, can hardly fail to succeed. On the other hand, he who breaks down elsewhere stands a poor chance there. The consumption of groceries, meats, horse-feed, clothing, boots, and the like, is enormous, owing to the vast influx of strangers, and the absence of economy in trifles, characteristic of all countries into which a great stream of wealth has been suddenly poured. Such articles usually sell at from fifteen to twenty-five per cent higher than in most country towns and villages, the prices depending somewhat on the means of transportation from the more prominent points. The cash principle is nearly universal. The apology made by the retail-trader for high prices is, that rents and horse-hire are oppressively high; that fuel and every other article *he* purchases costs two or three prices. In like manner, the owner of a house falls back upon the exorbitant sums charged for lots or ground-rents which are, he alleges, fully up to the New-York city standard; while labor and materials are so costly that it requires two or three times the sum to erect a store or dwelling that it formerly did.

Of hotels and boarding-house accommodations something has been said elsewhere. For transient guests the charges are nowhere below two dollars and fifty cents per day, and more frequently from three to feur dollars. Even at these metropolitan figures the stranger, who has been "taken in," cannot count upon a bed for himself, much less clean sheets and a separate apartment. At Shaffer's, for four dollars per day, he may perhaps get all these; and at Irvine, at a still more reasonable charge, with a manifest disposition to oblige. Everywhere the tables are abundantly spread with the substantials requisite for a good meal; but in a majority of places milk and the more delicate articles of fare are usually absent. The culinary arrangements are apt to be primitive, and cooks are of all imaginable grades. There is certainly little cleanliness in the eating, as in the sleeping apartments; but something must be pardoned to the genius of mud, as well as to the spirit of liberty. For permanent boarders the rate usually charged is ten dollars per week in hotels, and eight or nine dollars in boarding-houses.

If a man rates even moderately on that standard of measurement, "how to keep a hotel," and applies himself to his business, I hardly see how he can help getting rich in a few years, in spite of the extravagant rents and high prices of provisions, fuel, labor, etc. The rush of travel has hitherto been so immense that strangers have been completely at the mercy of landlords. The most tenderly reared and fastidious in taste have often been only too happy to secure six feet by two on a floor or a counter, wrapped in a buffalo-robe. No use in complaining about package, unventilated rooms, or dingy sheets; for a significant hint from the proprietor, that half a dozen others would have paid a premium for the same berth, invariably suppresses all uprisings of discontent. "Sir, we do the best we can for you. This is the oil country, you know."

But let the tide thitherward of excited folly decline, as it has already begun to decline, and hotel-keepers will find themselves competing for public "patronage," instead of patronizing the weary pilgrim by condescending to give him a coarse supper and a coarser bed for two prices. The owners of real estate will also begin to feel the uprisings of a sentiment akin to moderation. Lots for building purposes will be brought within the reach of industrious men, though they be not petroleum aristocrats; while those of more ambitious views may naturally aspire to the possession of *a whole acre*, to be cultivated for gardening purposes or devoted to pasture.

This brings me to consider the stolid stupidity of many

oil companies, in the discouragement they have offered to the cultivation of the soil by demanding prices as the rent of land such as no person could afford to pay. They have preferred to let thousands of acres lie idle to having them producing food for their workmen, or even hay for their horses. Their idea appears to have been that the property they have purchased as oil-lands should rent for only six month, at rates paying high interest on the investment! Short of such returns they would let those broad acres, which they were unable to bore, run idle. They have thus reduced an originally fertile country to the condition of a desert, depending upon remote localities for every measure of garden vegetables, every quart of milk, as well as the staple articles of food consumed. Hence, with all their high earnings, the Petrolians can seldom procure a glass of fresh milk or a mess of fresh vegetables. Hence, too, with abundance of land susceptible of cultivation, employés must pay two prices for board, which, in turn, reäcts upon the companies, compelling them to pay double wages, and reduce dividends in a corresponding degree. All this is attributable to the whirlwind of excitement, which has led all interests to despise the insignificant gains derivable from the soil, or from the practice of a prudent, far-sighted policy in procuring supplies at home for the rapidly increasing market. The numerous failures in proportion to the successes achieved there, an understanding by the public of the tricks and traps resorted to for the purpose of deceiving the unwary, must have the effect of bringing about (by necessity) a complete change of policy in this respect, on the part of interests which expect to survive. One or two Philadelphia concerns have already had the sagacity to see this and take action upon it. They offer such lands as are not likely to be taken up otherwise at moderate rents, for six months or a year, in the hope of inducing farmers and gardeners to migrate thither. But this is only a drop at the bucket. It is important that a *general* movement in the same direction shall take place, and that without delay. Migration on the part of the classes named must be systematized, and every reasonable inducement held out to them to take up their abode in the unoccupied lands. With a fair chance many hundreds would readily take hold there; and such men would constitute an efficient force of laborers to sink wells at a minimum price during the fall and winter months, as also to do hauling and other work at reasonable rates.

In fact, if petroleum is not to be brought down to, *but kept up with*, the general level of our great interests, the first proceeding in order is to strip it of the false glare in which it has hitherto been presented to the world; to inculcate the necessity of certain old-fashioned virtues like moderation, economy, prudence, forethought, public spirit, etc. After the money-making and money-losing fanaticism which has raged there for nearly half a dozen years, it may go hard with many to sit down to master afresh these rudiments of education; but the longer this task shall be put off, the worse for all concerned.

Owing to the impolicy noticed above, an evil of still greater magnitude—one which can hardly be measured in dollars and cents—has arisen. There is, in many places, scarcely a trace of what can be termed *a permanent element* in the population—certainly not outside of a few principal towns. In Oil City ninety-nine out of every hundred persons expect to leave it as soon as they have got "money enough." Hence, on the part of nearly every one, the understanding is to acquire wealth as rapidly as possible, without being too scrupulous as to the means; then clear out of the country, or "return to the States," as they term it. No impression could be more detrimental to the general well-being of a community than this. It bars all progress, except such as is concerned in momentarily inflating market-values. It brings the race-course with its excitement, but not the passable highway with its utility and comfort; the telegraph, with its cooked-up dispatches, not the decent sidewalk; the caravansary, not the well-kept hotel; the mountebank with his "gift enterprises," not the sound public teacher; the tribe of "Moses," not the class of honorable, public-spirited merchants; wild excitements, instead of the purer enjoyments becoming rational and accountable beings. It is an obstruction alike to good morals, pure religion, general education and refinement, as well as to public improvement. Where no person expects to remain, except for the briefest possible period, who feels interested in giving *tone* to a community? Who cares for its reputation outside, whether it be good or bad? Who will expend a dime in beautifying house, grounds, etc., adornments which add so much to the sweets of existence in beholders as well as owners? Who will oppose abuses, be they ever so monstrous, or institute reforms, be the necessity ever so urgent, when every step is certain to arouse opposition, and no one feels concerned about the distant future? The Spartans made their slaves drunk, in order to exhibit them in that condition to their children, and thereby fill their minds with disgust at the practice. If I wanted to impress on the mind of a youth the debasing effects of *selfishness*, I should transport him for a whole week to Oil City!

The social condition of Petrolia is thus in a measure that of California, for some years after "the gold fever" began to rage. Society remained utterly demoralized until the idea got into some men's heads that "the Golden State" might be made a desirable place for one's permanent abode. This revolution in idea was followed by the social uprising which overthrew the lawless element and established order, protection, quiet. What California has been made under the influence of a thought, Petrolia may become by the same instrumentality. Let this thought only take root there, and from it will spring up at once a beauteous growth of public spirit, generosity, kindness, accompanied by the march of improvement such as it has never yet witnessed. Until this idea has taken hold of the people, it is idle to expect aught but a continuation of the same recklessness, the same fluctuations in business, the same acquisition of princely fortunes by a few, at the expense of many and sad *miss*-fortunes on the part of the multitude. Petrolia will become a plague-spot on the score of manners, a great gambling-school, a camp of instruction for the whole country in falsehood and rascality. Principles which have lain latent in the human breast elsewhere, will there be "developed" equally with "oil territory;" and taking their places there as so many cardinal virtues, will come forth to walk the earth with unblushing front, and communicate to old and young the sad distemper.

However it may be with companies and individual speculators, he is a friend to the country and his race who exposes the abuses which have grown up in connection with the oil discovery; who proves that while there exist in Petrolia the conditions of an abiding prosperity, (the deposit extending to uplands as well as lowlands,) yet the average profits on extracting oil from the earth are nothing like so great as have been represented; that the agencies made use of to influence public opinion elsewhere are for the most part deceptive; that the men who have prostituted their tongues or their pens to induce strangers in such multitudes to invest their means, without any personal examination, in the business, belong to the class *knave* or the class *fool;* in a word, that it is by putting a stop to this temporary flush of prosperity, and turning men's attention to what is permanent, that the true well-being of that region is to be brought about. This I have at least endeavored to do.

It has been established, I think, that the river-system of that country has had nothing whatever to do with the original depositing of petroleum; consequently, that it may be expected with the same chance of success in the table-lands as along the bottoms, *plus* the cost and trouble of reaching and pumping it. It has been shown that in some valleys the supply of oil is fast becoming exhausted, whatever quantities may be brought up by boring new wells or deepening old ones. This is a painfully important truth, which, however, can no longer be concealed by the specious reasoning of scribes who have hired themselves out to puff general or individual interests. The hard fact of four-fifths of the old wells remaining idle year after year, in spite of repeated attempts to resuscitate them, also demonstrates that there is no chemical action going on which generates petroleum in large quantities out of its original constituents; but that the small quantities obtained are due to leakage, or some other equally simple agency.

It has been shown that many of the fortunes rapidly acquired in the oil regions by selling interests in productive wells or those expected to become such, are gained by unscrupulous lying and dishonorable stratagems, by which the uninitiated have been swindled and robbed, while in a state of mind approaching intoxication, from the "ardent spirits" distributed along the outposts of the country.

The actual yield of the producing wells has been ascertained or closely approximated to, and published for the first time. I feel confident that no well's production has been designedly misrepresented, and that the figures published as to what each was doing at the time of my visit, are not in one instance out of twenty far from the mark; that they are as often below it as above. From these *data* I have estimated the average production of the entire region; and though making no account of the loose estimates of others who were never on the ground, it is satisfactory to learn that the figures given here are nearly midway between the extremes. Estimates for working expenses, for replacing defunct concerns, excise duties, etc., have also been given; and a result on the score of profitableness arrived at, which will be found in accordance with common-sense and sound principle.

Finally, a number of practical suggestions have been offered, embodying not only the author's individual views, but those of experienced operators, whose knowledge and characters entitle them to a fair, full hearing on the part of residents, directors, legislators, and all others concerned in the permanent prosperity of the oil region. These various topics discussed, together with the processes of boring, refining, and repairing, the author would conclude by declaring once more that in dealing with every department of the subject his sole aim has been to arrive at and state "the truth, the whole truth, and nothing but the truth."

Made in the USA
Monee, IL
04 November 2026

67384998R00154